MINE CONTROL

Western Coal Leasing and Development

MINE CONTROL

Western Coal Leasing and Development

James S. Cannon

Research Associate: Bart Walker
Editor: Wendy C. Schwartz

Preface by Huey D. Johnson

COUNCIL ON ECONOMIC PRIORITIES
New York

Cover: Peg Averill

Production and Design: Susan Kent Cakars

James S. Cannon, who joined CEP in 1971, is the author of three previous CEP studies, *Environmental Steel, Environmental Steel/Update* and *Leased and Lost*. He has also written a handbook for environmental activists called *A Clear View* and co-authored a study of corporate alternative energy research and development programs, *Energy Futures*. He has written articles for many environmental, business, and energy publications and frequently lectures before university classes, citizen groups, and government agencies on environmental/energy topics. He is currently Research Director at Citizens for a Better Environment in San Francisco. He has an A.B. degree from Princeton University and an M.S. from the University of Pennsylvania.

International Standard Book Number: 0-87871-009-4
Library of Congress Catalog Card Number: 78-58006
© 1978 by the Council on Economic Priorities

Council on Economic Priorities
84 Fifth Ave., New York, N.Y. 10011
212/691-8550

Printed in the United States of America.
Printed on 70% recycled paper.

ACKNOWLEDGEMENTS

The author would like to express his gratitude to the many people whose assistance, guidance, advice and support made the completion of *Mine Control* possible. Cheryl Stevenson worked at every stage of the project, deftly molding vague ideas, characterless numbers and incomprehensible prose into what I hope is a reasonably coherent manuscript.

Individual chapters were expertly critiqued by Paul Robinson of the Southwest Research and Information Center, Carolyn Johnson of the Environmental Policy Institute, Tex Gary of the Powder River Basin Resource Council, Kathy Gramp of the Council on Energy Resource Tribes, Ron Rudolph of Friends of the Earth, Pat Sweeney of the Northern Plains Resource Council, and John Norton of the Dakota Resources Council.

I am indebted to the dozens of officials at the many state, local and federal government agencies who consented to interviews and who facilitated access to public records.

I would like to offer sincere appreciation to the staff members of Citizens for a Better Environment, in whose San Francisco office the final manuscript was prepared. I am particularly grateful for the trust and tolerance expressed by David Comey, CBE's President, and Phyllis Walters, its Director of Development.

Bob Faragasso of CBE and Stacie Linardakis at CEP deserve applause for their able typing of mountains of manuscript pages, as does Charma Pipersky who typed the Lease Catalogue, and Miranda Beeson who proofread it all.

I would like to give special thanks to Marty Teitel and Stephen K. Moody, Co-Directors of the West Coast office, Nancy Nation, Mary Richardson and Deborah Doyle, and to CEP intern, Audrey Kasun (University of California, Berkeley), all of whom worked in San Francisco to get the project off the ground.

Bart Walker ably worked on every phase of the project. Wendy Schwartz reviewed and skillfully edited the study. Alice Tepper Marlin, CEP Executive Director, enthusiastically supported the project idea and carefully reviewed all drafts of the manuscript. Finally, Susan Kent Cakars spent long hours turning the manuscript into an attractive book.

I would also like to thank CEP Board Members Maryanne Mott Meynet and Jake Kittle for their encouragement and advice on the project, and the entire CEP Board for its support.

For their critical financial support of this project, CEP would like to express its deep appreciation to the New York Community Trust, the Shalan Foundation, the Nu Lambda Foundation, and the many individuals and foundations that contributed to CEP's general funds.

James S. Cannon
October 1978

TABLE OF CONTENTS

PREFACE

Coal development is one of the most important resource questions the U. S. has faced in its history. If we don't act wisely, our culture could become like ancient Rome—extinct. Natural resources are the basis of a nation's vigor, and lessons from history, such as ancient Roman wine presses found in the Sahara Desert, show how cultures declined when resource management was ignored.

The nation's vast coal resources offer a means of orderly transition from an economy based on maximum exploitation of limited resources to one based on sustained yield utilization of renewable resources. We know, now, that resources are not unlimited; but we did not learn that lesson until we were to the point of degrading our forests, fisheries, air, water, soil and oil supply. Indeed, current threats to our society, like inflation, are a powerful hangover from that grand binge when we ignored the earth as a basis of wealth.

At the moment, we still have a vast, available resource—coal, and we have advantages over the Romans. We can look at their errors, and we have more new information on how to manage for survival. The questions are: Can we apply what we know? Can we manage that coal resource logically?

Though I'm optimistic we *can* manage well, I am not so optimistic we *will*. This book factually defines the issue and courageously prescribes changes that will provide a basis for management.

The first challenge is ownership. The public has abdicated control over the resources found on our lands. We the people, each of us, own 2.7 acres of federal land, but the leasing program on this land is, as this book elegantly and precisely demonstrates, a mess. It was designed in the 1800s for the horse and buggy era—a time when resources were viewed as unlimited, the land was free, and the rule was to get it, develop it and sell it. Like the Indians who sold Manhattan Island for a few glass beads, we've given up our public coal lands containing both the wealth of kings and the energy for our future.

A fair return to the public for the right to mine and sell minerals on public land was not considered when present leasing practices were created. Now it's beginning to be. Most states realize that nonrenewable resources, like coal and oil, when used bring in no more revenue.

We need to plan. Alaska's futuristic permanent fund, a royalty per barrel of oil, expected eventually to total $100 billion and allowing annual expenditures of interest only, is a sound example of planning for the future. Coal can be used the same way. By saving from the boom-time shares, we can use a nonrenewable resource to create a permanent fund to keep our public services humming later.

And whether we get a fairer shake in value, we the public must at least get a better basis for professional management. The scale of coal use is an awesome challenge. For example, we must study the threats of pollution that require technology which is not quite perfected. We must tackle the mind-boggling problem of coal transportation. We must reckon with the atmospheric effects of CO_2 released from coal combustion. And, we must resolve the conflict of the competition for other already used resources, such as whole rivers that would be diverted from farming and recreational uses.

In California we have struggled, with considerable success, with air quality and other resource–environment–energy problems. The confusing public coal land use policies remain a little understood but major obstacle in determining the role of domestic coal in California's energy future.

Since the state's future use of domestic coal will be the single most important event in the future of western coal, it's important that some land use policy problems be solved. This book makes a significant contribution by exposing the problems involved in coal development.

Huey D. Johnson
Secretary for Resources
State of California

INTRODUCTION

The West is once again attracting settlers. They are coming this time not for the land, but for the coal which lies beneath it. Their tools are not the plow or the lasso, but machines capable of mining small mountains with each bite. Their weapons are not guns, but the multibillion dollar force of the world's largest corporations. Their government allies are not the cavalry, but land management agencies. Their moral justification is not "manifest destiny," but the "moral equivalent of war," for which Jimmy Carter says the "most formidable defense weapon in our arsenal is coal." Their legal rights are not simply taken, but are given in the form of coal leases issued by state and federal governments. These settlers, like the ones before them, promise to change the face of the West.

President Carter's 1977 Energy Message called for a doubling of national coal production by 1985. He is not the first national leader to call for such a rapid expansion. Gerald Ford's 1975 National Energy Plan included the same objective, as did Richard Nixon's Project Independence, his response to the 1973 Arab Oil Embargo. None of these administrations formulated this policy, however. Each merely elevated to national policy the plans of the coal industry and the electric utilities.

Throughout the 1970s coal production in the country has indeed been expanding. The opening of the western coal fields has been the single most important event fueling this increase. According to every coal production forecast of the decade, western coal will spark national coal growth for years to come. How much coal the nation consumes over the next several decades will be determined more by western coal leasing policies than by any other single factor.

Western coal accounted for 18% of national output in 1977, doubling over four years. More than half of the nation's coal is located in western deposits, and about 85% of this western coal is on public or Indian land. The federal government clearly overshadows both state and Indian influence. In no major western coal-bearing state does the federal government control less than 25% of the coal containing acreage and its overall control includes about 65% of all western coal acreage.

Western coal development means a relative decrease in mining on private property and an increase in mining on land belonging to the American public and Native Americans. Whatever development occurs in the West is made possible solely because western coal's present owners, state and federal governments and Indian tribes, have agreed to lease coal-bearing land to private developers.

The emergence of the West as a center for coal production marks a shift from underground to strip mining technology and reliance on nonunion rather than union labor. It means fewer hazardous underground mining jobs, but greater environmental devastation from strip mining. Western coal mining and industrial expansion stand in stark contrast to the existing agrarian industries and local lifestyles. They also stand for new jobs and economic opportunities for the West, especially if coal conversion technologies such as gasification become commercial. The job gains, of course, occur at the expense of jobs in Appalachian coal regions, and the benefits of western coal development are at least partially offset by the ecological, social and economic disruption.

The fast growth of western coal development has meant new members for the coal club: they are the oil giants, energy conglomerates, utilities, and construction companies. The traditional coal industry—eastern-based undiversified mining companies—is being squeezed out by the better capitalized, diversified corporations which are amassing huge coal reserves in the West. Small Appalachian mines, traditionally the backbone of the U.S. coal industry, are losing ground, and losing it fast, to the massive strip mining operations in the West. Increasingly coal mining, like many other industries, is becoming one of the many activities of integrated and diversified corporations.

The Council on Economic Priorities has been studying western coal development since the early seventies. In 1974 CEP published *Leased and Lost,* the first independent study of the U.S. Department of Interior's leasing programs for public and Indian coal land in the West. *Leased and Lost* covered seven states: Arizona, Colorado, Montana, New Mexico, North Dakota, Utah, and Wyoming. It helped spawn the nationwide debate about whether or not unrestricted development of our vast western coal reserves is in the public's interest, and it helped form the basis for the Federal Coal Leasing Amendments Act of 1975.

This new study, *Mine Control,* is both an update and an extension of CEP's research on western coal development. It chronicles the history of the federal leasing programs between 1973 and 1977, updating the *Leased and Lost* findings. It also includes a new in-depth analysis of leasing programs operated by state governments in six western coal-bearing states (Arizona has been excluded because it has issued no leases). It evaluates the current status of coal development in each state and ascertains each state's formal or informal coal policy. CEP has examined each individual federal, state and Indian coal lease and has catalogued basic data about each lease in an appendix volume.

2

A rush for coal leases preceded both the present boom in coal development and the establishment of the national energy policy justifying it. For nearly 20 years, the groundwork for this development has been built in the offices of state-land leasing offices, in county land offices, at Indian Tribal Councils, and in the regional offices of the U.S. Bureau of Land Management and the Bureau of Indian Affairs. It is there that the terms of several thousand lease contracts, each of which transferred ownership of some western coal from the public or Indian domains to the private sector, were set. These leases, issued long before coal prices justified start-up of mines on the land they covered, are now the pillars supporting the coal boom. This book is about them.

SUMMARY OF FINDINGS

KEY FINDINGS

1. Western coal mining is booming. The amount of coal produced on federal land tripled between 1973 and 1977, while output from Indian land doubled. Eight of the 10 largest mines in the country are now located in the West.

2. Approximately 16 billion tons of federal coal reserves are under lease, 300 times the 1977 output from federal land and 24 times the 1977 total national production. Despite the federal "moratorium" on coal leasing, which has been in effect in various forms since 1971, 13 new leases have been issued since 1973 and eight have been relinquished.

3. Only two of the 10 outstanding Indian coal leases have been successfully renegotiated since a 1974 Department of Interior ruling essentially invalidated all Indian contracts. The others remain in legal limbo, though coal mining has continued unabated. The U.S. Bureau of Indian Affairs has done little to improve the advisory services it provides to Indian tribes on coal leasing issues. The Indian tribes contacted by CEP continue to be highly critical of the government program and are moving independently to manage their coal land, protect their lifestyles and correct the travesties of past leasing errors.

4. Coal developers have turned their attention in recent years to state-owned coal land. Six western state governments have issued 2,553 leases, half in the past five years, covering three times the land under federal lease and nine times the acreage covered in Indian contracts.

5. Lease issuance procedures have amounted to coal giveaways. Federal and state governments and Indian tribes have repeatedly issued leases at the first sign of interest, long before the demand for coal became competitive. The fee is usually just a few dollars per acre.

6. State leasing agencies have repeatedly given away coal rights for far less than they are worth, in part because of confusing directives contained in an archaic land management program termed by CEP, the "beneficiary institution system." It severely limits the involvement of leasing agencies in land or resource management decision making and allocates all revenue from state land to 82 specific institutions.

7. Most western state leasing programs are in desperate need of reform. Two states, North Dakota and Montana, have taken important steps to improve their leasing programs. Colorado and New Mexico have also instituted reforms, but their actions are less extensive and effective. Utah and Wyoming continue to operate completely inadequate leasing programs and have expressed no intention of improving them.

8. Lease speculation is rampant. Western coal leases are sold by the lessees in a process called "assignment." Over one quarter of all federal leases have changed hands since 1973, and 34% of all state leases have been assigned at least once. The assignor keeps all income from assignments.

9. Because a majority of leases are being held for speculation, only 14% of all federal leases and a miniscule 0.7% of state leases are now in production. Thus, enormous coal expansion in the West could easily be supported without additional leasing.

10. Rent, royalties, and advanced royalties charged to leaseholders are pitifully low, contributing to the speculative holding of leases and to a wholly inadequate return to public and Indian treasuries.

11. The five largest federal leaseholders control about 31% of federal coal leases and the top five state leaseholders control an average of 45% of the leases in each state. Peabody Coal controls more than one third of all Indian land under lease.

12. Exxon is the largest holder of state, federal and Indian leases in the West. Oil companies in general play a dominant role in western leasing.

13. Washington has been the site of considerable activity on federal coal issues. New legislation, regulations, studies and Environmental Impact Statements have appeared since 1973. These actions, though encouraging, are incomplete and somewhat misdirected. The Department of Interior has failed to establish a position on some key issues, such as the disposition of preference rights lease applications or the resolution of conflicts between surface owners and coal developers on leased land. Its efforts have been directed toward developing a leasing program for the future, but it has not dealt effectively with the problems created in the past. CEP recommends that Interior focus on a coal development program for the substantial acreage already under lease.

14. CEP further recommends that the beneficiary institution system be amended, through Congressional legislation and revision of state consti-

tutions, to require states to consider resource management, both in the short and long term, as well as income maximization, in managing their coal lands.

* * * * * *

The single most significant observation made by CEP while preparing *Mine Control* is that coal development in the western United States is proceeding at a ferocious pace. This study discredits the industry claim that the federal government's leasing "moratorium" (in effect in various forms since 1971) has put a lid on coal development, thereby endangering our ability to meet President Carter's goal of doubling 1976 coal production by 1985.

Coal production on federal land in the West is increasing, notwithstanding the "moratorium." The total output of the 67 active federal leases in 1977 was 52,443,593 tons, a 241% increase in coal output since 1973, when CEP did its first study of coal leasing. Total national production jumped a mere 14% in the same time frame. In 1973 federal coal lands under lease contributed 25% of the western coal market; in 1977 the figure stood at 42%. The dramatic increase in coal production on federal land points to the emergence of the federal government as the most significant single entity affecting coal development in this country.

Coal production on Indian land is also on the increase. Output has more than doubled since 1973, to 22,958,047 tons in 1977.

Mining operations on state land yielded 6,651,345 tons in 1977. This relatively low figure reflects both the newness of coal development at the state level, and the high degree of lease speculation there.

Western coal mining is dominated by strip mining operations built on a scale unheard of on the eastern coal fields. Eight of the ten largest U.S. coal mines in 1977, including the top six, are located in the West on public or Indian land. All eight utilize strip mining technology.

The pace of development is not expected to diminish. Bituminous Coal Research, Inc., a research arm of the coal industry, predicts that two-thirds of the coal expansion this country will experience by 1985 will occur in the West, that 47 existing western mines will be expanded, and that 97 new ones will be opened.

EXTENT OF LEASING

The coal production boom was preceded by tremendous leasing activity in the West which continues today. As of 1977, there were 468 valid federal coal leases in the states included in the CEP survey, which covered over 90% of all federal coal leases nationwide. Federal coal leases in these states now cover

703,193 acres. Approximately 16 billion tons of coal reserves are leased, more than 300 times the 1977 output from federal land and 24 times the 1977 total national production. Since 1973, 13 new leases have been issued and eight have been relinquished during the Department of Interior's various leasing "moratoria."

There are presently ten outstanding Indian coal leases in the states surveyed by CEP, one fewer than in 1973. Peabody Coal relinquished its nonproducing lease on the Southern Ute reservation in 1975. The ten leases cover only 239,902 acres, but this land supports three of the ten largest mines in the country. All Indian leases were invalidated in mid-1974 by the Department of Interior, but after four years, new contracts have been signed for only two leases; the others remain in a state of legal limbo. However, coal mining activity has proceeded unabated.

Coal developers have historically turned to the huge unbroken blocks of consolidated federal and Indian coal land instead of the small scattered sections of state-owned land. This situation has changed dramatically since government reform movements began in the early 70s. Coal brokers have turned their attention to state land. At the end of 1977 there were 2,193,630 acres of state-owned land included in 2,553 valid state-issued coal leases. This is over three times the amount of land under federal lease and nine times the land covered in Indian leases. Exactly half of all state leases have been issued since the beginning of 1973.

LEASE ISSUANCE

The coal leasing programs operating in the West fail to guide and regulate coal development in the public interest. A major reason for this shortcoming is that leasing agencies do not act as either land and resource managers or energy planners. Resource management is desperately needed in the West. Since coal land leasing passes public resources into private hands, it is a logical and effective place to implement land and energy use decisions. Yet the leasing agencies generally leave the task of energy planning to others, content to act as accountants.

All coal leases in the West have been issued according to one of five methods—"open land," "preference right," "competition," "negotiation" and "segregation." The common bonds linking these systems are their consistent failure to regulate lease issuance in any concrete or logical manner, and their failure to generate significant financial return to public or Indian treasuries.

Open land leasing is the system most conducive to the random giveaway of public resources. Approximately 85% of all state leases have been issued by the open land method, garnering for the states a mere $47,653, an average of 2¢ an acre. Although most states have recently abandoned it, open land leasing remains the principal lease issuance procedure in Utah and Wyoming.

Preference right leasing, another noncompetitive system, has garnered an average of only 22¢ for each of the 9,085 acres of state land leased under this program. Nearly half of the 468 outstanding federal coal leases have been issued under this system, earning a total $2,100, just a fraction of a cent for each acre.

CEP found that three of the 13 new federal leases issued since 1973 employed the preference right system. They include the second and the sixth largest leases ever issued. The public received a total of $30 from the issuance of these three leases which are known to cover millions of tons of coal.

The preference right leasing program was discontinued by federal legislation in 1976, but approximately 180 preference right lease applications are still pending from permits issued before the program was abolished. If these are allowed to be converted into leases—and recent court decisions suggest they might be—an additional 476,000 acres covering 12 billion tons of coal, nearly as much coal as is currently under lease, will fall into the hands of the private sector for a total payment of $1,800. At today's prices, 12 billion tons of coal would sell for at least $120 billion.

Approximately 13% of all state leases, slightly over half of all federal leases, and exactly 50% of Indian leases, have been issued according to various "competitive" leasing systems. State leasing agencies have accepted an average winning bid of a mere $6.06 per acre at 333 lease sales. Indian tribes have garnered an average of $10.32 per acre. Buoyed by a few recent sales which attracted several hundred dollars per acre, the average winning bid by the federal government from its 239 competitive lease sales now stands at $35.44. While relatively high compared to state and Indian collections, the average price per acre still equals the selling price of just two or three tons of coal. Only one bidder appeared at 131, or 72%, of the state land competitive lease sales for which complete data is on record. Just one bidder appeared at 141, or 59%, of all federal competitive lease sales, paying an average of only $5.39 per acre. In contrast, the average winning bid at federal sales attracting four or more bidders was $179.27 per acre. Unfortunately, these truly competitive sales account for only 6% of all federal auctions to date.

Negotiated leasing was used for only five Indian and four North Dakota state leases. Although this method has the greatest flexibility and requires the largest involvement of leasing agencies, to date it has not been an improvement over other methods, due to the failure of the agencies to press for better terms.

Finally, 29 state leases and 19 federal leases have been created by segregation, the division of an existing lease into two distinct entities. Segregation fees undoubtedly do not cover the costs of the paperwork involved in preparing new lease contracts: states have received a total of $230 from segregations to date; the U.S. Treasury, $190.

State and federal leasing programs are simply giveaways. The average state lessee has paid a mere 55¢ per acre, while federal lessees have been pressed for only $17.79 per acre. And the Indian tribes have been advised by the U.S.

Bureau of Indian Affairs to accept the ridiculously low terms offered by coal developers. By any standards, it is clear that "fair market value" is not being obtained from lease issuance systems. Coal sells for about $10 per ton, and a single acre of land typically covers 50,000 tons. Thus, an investment of a few dollars per acre for the lease brings the winning bidder more than half a million dollars worth of coal. In 1976 the coal industry's operating profit margin was a high 17.7%.

SPECULATION

The issuance of coal leases for paltry sums, before the existence of competitive interest by coal developers, has set the stage for one of the most disturbing phenomena of current western coal development—the selling of public land and resources for private profit. The present leasing systems allow lessees to sell their coal leases to another party by a process called "assignment," at any time and without any additional payment to the government. The assignor, who may never have had any intention of extracting coal from leased land, is permitted to keep any profit obtained from the transaction.

There is no single aspect of the present coal leasing programs in the West which fosters as much speculation, land fraud or confusion among national energy planners as do lease assignments. More than one-quarter of outstanding federal leases have changed hands through assignment since 1973. Although the average state lease is much more recent than federal leases, 34% of all state leases have already been assigned at least once.

The fast pace of lease assignments in the West is an indicator of speculation. Lack of production on leased coal land is another persuasive indicator. In 1977, just 14% of all federal coal leases were in production. This 86% inactivity rate and the lack of any firm development plans for about half of the coal leases (according to Department of Interior studies) testifies to the enormous coal expansion which the West could support without additional leasing.

There is even less production on state leased land. Only 19 leases, or a miniscule 0.7% of all state leases, yielded coal in 1977. There is only one state, North Dakota, where more than 2% of existing leases are now in production.

LEASE OWNERSHIP

While there are over 500 lessees in the West, CEP has found a large and increasing consolidation of leased land in the hands of a few corporations. Roughly 31% of federal leased land is controlled by the five largest leaseholders. The top five holders of state leases control an average of 45% of leased land in each state. Peabody Coal alone controls over one-third of all Indian leased coal land.

Oil companies play a dominant role in western coal leasing. Exxon, through its subsidiary, Carter Oil, emerged as the largest overall holder of federal, state and Indian leases in the West; it has acquired over 181,355 acres. Continental Oil's subsidiary, Consolidation Coal, is the fourth largest, and Ark Land, a subsidiary of Arch Minerals, which in turn is a joint venture of Ashland Oil and Hunt Enterprises, is number five.

Peabody Coal, now owned by a consortium including Bechtel, Boeing and the Equitable Life Assurance Society, is the number two leaseholder. Rounding out the list, as the third largest, is Mapco, one of the fastest growing resource conglomerates in the country.

Energy conglomerates are making widespread use of the assignment market to increase their dominance. Gulf Oil, for example, acquired seven federal leases by assignment from four different leaseholders since 1973. Over the same period, American Electric Power's subsidiary, Franklin Real Estate, acquired 11.

The continuing increase in the control of leases exercised by energy conglomerates, such as Sun Oil, Arco, and Continental, is reflected in the increase in coal production from mines owned and operated by them. The Bituminous Coal Research, Inc. estimates that oil companies will mine approximately 61% of all the coal now under contract to be supplied from western mines.

REVENUE

Rent, the yearly "carrying charge" for holding onto a coal lease, must be paid by every leaseholder until coal mining begins. Rental collections are pitifully low and contribute to the speculative holding of land. Lessees paid the federal government and Indian tribes approximately $1 per acre in rent in 1977. The six state governments collected an average of just 41¢ per acre in rent from state land lessees in Fiscal 1977.

Some lessees are required to pay advanced royalties if mining operations do not begin on leased land within a specified time frame. Advanced royalties are intended to provide an incentive for leaseholders to begin mining operations at an early date, but CEP has found that they do not accomplish this goal because they are set so low. In Fiscal 1977, state governments collected an average of 61¢ advanced royalties per acre leased.

Virtually all outstanding federal and Indian leases use a fixed royalty payment for every ton of coal mined. The going rate is 10-30¢ per ton. The federal government collected average royalties of 19¢ per ton of coal mined in 1977. With an average selling price of coal at $10 per ton in the West, this represents just 1.9% of the market value of coal mined. The Federal Coal Leasing Amendments Act of 1975 requires that royalties be at least 12.5% of the selling price of coal.

Since 1971, government policy has supported a switch from fixed to variable royalties, which require lessees to pay a royalty based on a percentage of the selling price of coal. The Department of Interior has the right to change royalty rates before renewing a lease contract. However, of 35 federal coal leases that have come up for renewal since the royalty policy was changed in 1971, not one has had its royalty rate adjusted. These leases include eight active mines which yielded about 4.1 million tons of coal in 1977. CEP calculates that, by not upgrading the royalty rates, the federal government lost about $1 in royalties for each ton, or more than $4 million in 1977 alone.

The tribes housing the five active coal mines on Indian leases are faring a little better. They received an average of 27¢ per ton of coal mined from their land.

The lease terms of many recently issued state coal leases include variable royalties ranging from 5% to 12.5% of the value of coal mined. In 1977, however, the state leasing agencies collected even lower average royalties than the federal government or Indian tribes, just 18¢ per ton of coal mined. This situation occurs because most present mining operations exist on the older state leases which have fixed royalties.

STATE LEASING

The tragic mismanagement of land and resource development responsibilities so characteristic of western leasing programs in general is most graphically demonstrated at the state level. More distressing still, most state leasing activity has occurred after the abuses of the federal program had already been exposed and the need for land management reform demonstrated.

State leasing authorities occupy the bottom rung of the western mineral development government hierarchy and are often ignored by other planning agencies. Partly in an attempt to compensate for, or bypass, the state leasing apparatus, most state governments are moving to enact higher coal severance taxes. Among the six states examined, only Utah has no such tax, although Colorado and New Mexico do not take a very big bite from coal developers revenues. Montana's tax of nearly 30% of the value of coal mined within the state is currently the highest in the West but is being challenged by coal miners and utilities as confiscatory.

Severance taxes, however, are essentially a tool available to a state to regulate the pace of coal development, and to raise revenue. They apply indiscriminately to all coal in a state, whether it is mined from federal, state or private land. Hence severance taxes do not substitute for state coal leasing policies, which should be directed toward sound resource management as well as obtaining a fair price for the public for the loss of one of its assets. Since all severance taxes are spent solely by the states which levy them, they cannot replace the revenue generating components of the federal leasing program, which should serve all citizens of the United States.

The failure of the states to effectively administer their own leasing programs can be traced to an inflexible and archaic program, which CEP terms the "beneficiary institutions system." When each state entered the Union, the U.S. Congress assigned a set of institutions as sole recipients of revenue from particular state lands. Although the common schools are the chief recipients of revenue from activities on state land, CEP has found a total of 82 institutions in the six states—including prisons, hospitals and insane asylums—which are the direct beneficiaries of coal leasing activity. It is the singular objective of state leasing programs to maximize the financial return to these institutions. Worse still, they have failed to do even this, taking any immediate return rather than managing the land to maximize income over the long term. All of the agencies surveyed by CEP historically have equated their responsibility to maximize revenue from state land with an obligation to maximize commercial activity on state land. This erroneous judgement has led to the indiscriminate and wholesale leasing of vast tracts of state land, without responding to either competitive interest in the land or market value for the coal, conditions conducive to charging much larger amounts for leases.

The beneficiary institution program is anathema to sound resource management. By tying coal development on state land solely to the budgetary interests of state institutions, the state leasing agencies are prohibited from developing and implementing sound resource management. While few people would question the social justifications for most of the beneficiary institutions, the system should be amended at least to permit, indeed to promote, the establishment of resource and land use expertise within state leasing agencies and to enable them to institute resource development consideration as a key criterion.

Most states have suspended leasing for short periods this decade to examine their leasing programs. But CEP concludes that only North Dakota, and to a lesser extent Montana, have substantially improved their programs from the abysmally designed and administered systems historically employed. North Dakota's new official leasing policy "to maximize values, not revenues" is exemplary. Less sweeping changes have been made by New Mexico and Colorado.

Wyoming and Utah remain as blemishes on the past, the present and the future records of state leasing. They operate the worst systems and have made no attempts to correct them. Furthermore, these states have leased virtually all their coal land. Neither agency expressed to CEP any intention of reforming its program.

State governments are also calling for increases in federal aid, claiming that the adverse impact of coal development can be traced directly to a maladministered federal leasing program. The efforts of Montana and North Dakota notwithstanding, the simple fact exists that state governments have done little to "keep their own houses in order" through competent management of their own coal land. Until widespread reforms permeate state leasing programs, the cries of state governments for help must be viewed with skepticism.

INDIAN LEASING

Native American activists focused attention on the weaknesses of the Bureau of Indian Affairs in the early 1970s. Under heavy public pressure, the government promised to be more responsive to Indian needs. BIA has failed in the arena of coal development: the regional BIA offices have not upgraded their services to the tribes on coal development questions. Few new federal regulations have been promulgated and no new Congressional legislation has been proposed. The Indian tribes contacted by CEP continue to be highly critical of the Bureau's performance.

The five Indian tribes which are engaged in leasing coal are now acting on their own to control coal development on their reservations. In 1973, the Northern Cheyenne petitioned the Department of Interior to have all coal leases and permits on their reservation canceled because of egregious breaches of leasing codes made by the BIA when the contracts were negotiated. The tribe was victorious, setting the stage for the renegotiation or cancellation of all existing Indian leases. In the past four years, however, only two leases have been successfully renegotiated, while the rest have been embroiled in legal actions, negotiations, and intra-tribal debates. Each tribe, however, is finally moving on its own for the first time toward a concrete coal development policy, which could have broad implications for U.S. coal development.

Individual tribes are moving in other ways to rectify past travesties and to protect their lifestyles in the future. The Navajo, for example, have enacted three new taxes on coal developers. The Northern Cheyenne successfully petitioned the U.S. Environmental Protection Agency in 1976 to classify their reservation as a Class I area of pristine air quality, thereby blocking the construction of two power plants near the reservation.

Western Indian tribes have recently taken the first major step toward inter-tribe cooperation. In 1975, 22 tribes formed the Council of Energy Resource Tribes (CERT). With a Washington lobbying arm and a Colorado research division, CERT aims to study Indian resources and to provide tribes access to a pool of resource development expertise.

FEDERAL LEASING

Coal leasing issues have generated considerable activity in Washington. In 1976 Congress passed the Mineral Leasing Act Amendments which incorporated the bulk of the *Leased and Lost* recommendations and received the strong support of CEP. Several major environmental lawsuits have been resolved. The Carter Administration has established two new divisions within the Department of Interior to deal with leasing issues. A dozen or so new improved leasing regulations have been promulgated. Scores of studies have meticulously dissected the existing leasing program, as well as the many hypothetical programs which could be implemented. Approximately 10 Environmental Impact Statements are in various stages of completion.

Encouraging as these actions have been, especially in light of the ennui of Carter's predecessors, the current government program is incomplete and misguided in direction. Interior still has not come to grips with several key issues. For example, it has repeatedly waffled on the question of leasing subsurface mineral rights where the surface is owned by a private party, a situation on about 39% of federal land leased to date. The Department of Interior has not adopted a program to guide the resolution of land use conflicts between surface owners and subsurface lessees. Furthermore, it has not established a consistent position on future leasing in such cases.

Another issue lacking a clear government stance is the disposition of pending preference right lease applications. Many citizen groups have urged the Department to seek to cancel these lease applications, and a recent court decision suggests that the government could have broad powers to limit the development potential of these lands through strict lease terms. The government, however, has not yet stated a policy. Are further coal leases essential to future development? Interior has not answered this question and cannot answer it at least until the preference right leasing issue is resolved.

A major failure of the Department of Interior's leasing program is that current leasing reforms do not generally affect the administration of existing leases. The government's primary objective is to develop a leasing program for new leases. By essentially ignoring past actions, the government has turned its back on the continuing adverse effect of its inept and outdated 1920 program.

With nearly three-quarters of a million acres of coal land under lease, covering 16 billion tons of coal, the government need not lease more land. Rather, it should concentrate on developing a program to mine the resources already leased. Land leased in environmentally unacceptable regions should be exchanged for land with a favorable environmental rating from the government. Lease contracts could be promptly renegotiated at the conclusion of their 20-year terms to include higher royalties and rent and diligent production clauses. Mining plans should incorporate rigid environmental protection measures or not be approved. Speculation should be minimized by stringent production requirements and by the abolition of the lessees' right to assign public land leases indiscriminately. All land should revert to the public domain when the leaseholder is unable to develop a mining plan for the land.

Future federal leasing actions should be viewed in the context of the existing program, recognizing that no new leasing is necessary or acceptable until the present damage is repaired. The U.S. Environmental Protection Agency has established a national policy, called the tradeoff program, applicable to areas where air is more polluted than federal standards deem necessary to protect human health. Under this program, a new industry must reduce air pollution from other sources in an area at least as much as its operation will contribute to air pollution. Tradeoffs will be required until national air quality goals are achieved in that area. The Department of Interior would be well advised to establish a similar perspective to guide its leasing program.

OVERVIEWS

1. FEDERAL LEASING

INTRODUCTION

The U.S. federal government is the largest single coal owner in the United States. Nearly 40% of the nation's 1.5 trillion tons of recoverable coal reserves lie buried under federal land. Virtually all of this coal is found west of the Mississippi, where the federal government owns 65% of all coal lands. In no state included in the 1,000-mile-wide belt of the six states CEP surveyed—Colorado, Montana, New Mexico, North Dakota, Utah and Wyoming—does the federal government control less than 25% of the coal land. In Montana, approximately 75% of all coal-bearing acreage is under federal domination, while in Utah the figure is a staggering 82% (see Table 1-1). Especially in the Northern Plains, where complicated land ownership patterns affect state, private and railroad land, the federal government owns outright more prime coal areas than any other land owner.

Development of coal reserves in the public domain is clearly pivotal to western coal expansion. The availability of western coal hinges largely on a 58-year-old government leasing program operated by the U.S. Department of Interior in accordance with the Mineral Leasing Act of 1920 as amended. CEP has been a long-time critic of Interior's leasing program. Our 1974 study, *Leased and Lost,* found that mismanagement by the Department of Interior had consistently made the worst of a bad situation. The Mineral Leasing Act, originally intended to encourage exploration and settlement in the West, did not provide directives for orderly development of public resources at fair market prices. Acting without a strong Congressional directive, Interior willingly and haphazardly transferred three quarters of a million acres of land to the private sector without regard for land use planning, without receipt of fair market value, without assurance of orderly development, and without taking adequate steps to protect the environment. The major findings of *Leased and Lost* are summarized in Table 1-2.

TABLE 1-1

FEDERAL COAL LANDS

	Federal Coal Land (million acres)	% of State's Coal Land	Total Demonstrated Coal Reserves on Federal Land (million tons)
Colorado	8.7	53	14,870
Montana	24.6	75	9,488
New Mexico	5.5	59	108,396
North Dakota	5.6	25	16,003
Utah	4.1	82	4,042
Wyoming	11.8	65	53,366
TOTAL	60.3	65	206,165

TABLE 1-2

SUMMARY OF FINDINGS OF *LEASED AND LOST*

The Federal Coal Leasing Program has been the keystone of coal development in the western United States. In 1973, CEP reviewed the 474 coal leases covering federal and Indian land in seven states—Arizona, Colorado, Montana, New Mexico, North Dakota, Utah, and Wyoming—issued by the Department of Interior.

Our major findings were as follows:

Legislation: The Mineral Leasing Act of 1920 and the Omnibus Tribal Leasing Act of 1938, which authorized the leasing of public and Indian mineral rights, have not worked in the public interest. Neither law required resource use or land use planning. Neither compelled Interior to implement directives for orderly development of federal and Indian resources at a fair market price.

Extent of Leasing: Over 15 billion tons of recoverable federal coal and five billion tons of Indian coal were under lease—35 times the amount of coal produced in the U.S. in 1973.

Concentration of Holdings: The top 15 of 144 leaseholders controlled 70% of all land under lease. Five oil companies and three electrical utilities were among the top 15.

Speculation: Only 52, or 11% of the 474 leases were producing. Three hundred and twenty-one leases had never produced a single ton of coal. In its 54 years of operation, the leasing program had contributed less than 1% of the nation's coal production.

Lease Terms: Every lease had been issued at industry's request rather than as a result of Interior's determination that there was a market demand for coal. Two hundred and forty-seven of the leases had been issued at competitive lease sales, but 171 were granted without competition—since one or no bidders appeared—for an average of $2.87 an acre. Another 210 leases were granted by the preference right method which returns no revenue except the $10 filing fee. Rental rates seldom surpassed $1 per acre. Royalty rates also were extremely low, averaging 12.5¢ per ton on federal land and 15.8¢ per ton on Indian land. Over the course of the program, royalty rates had risen 75% while the price of coal had more than doubled.

Reclamation: Reclamation—the restoration of the original contours of the land—had occured on only half the 6,515 federal acres that had been strip mined.

The preparation and publication of *Leased and Lost* coincided with an increasing concern within the federal government about the mismanagement and inadequacies of the then prevailing leasing system. Partly on the strength of its own analysis and partly under the heavy pressure created by the exposure of information contained in such publications as *Leased and Lost,* the Department of Interior instituted and maintained a leasing "moratorium" for most of this decade. The moratorium, originally scheduled for three years, will continue at least until mid-1980.

CEP has continued to observe, chronicle and analyze the federal leasing program. The moratorium notwithstanding, there has been considerable activity in the publicly-owned coal fields in the West, in the Congress, in the federal bureaucracies, and in the federal courts in the East. Change is occurring at a ferocious pace, and legislative improvements have been substantial. But CEP believes that, despite many individual changes for the better, the federal government has failed to establish an overall national coal leasing policy. Operating without any clear goals or objectives for western coal development on public land, the Department of Interior is no more equipped in 1978 to implement an adequate coal leasing program than it was in 1974.

MORATORIUM

The word most commonly used in connection with the federal coal leasing program in the 1970s is "moratorium." Despite its widespread use, the word means different things to the various persons and institutions interested in western coal development. Coal industry representatives consider the moratorium a virtual lid on coal leasing and development clamped down upon them by the Department of Interior. Interior considers it a necessary transition between the old, inadequate program and a newer, more sound one. CEP has found both of these explanations insufficient.

The moratorium on coal leasing in the 1970s represents a series of Departmental policies, each with varying effects on coal leasing activity, rather than one single Interior pronouncement. When a November 1970 study by the Bureau of Land Management (BLM) revealed that the massive upsurge in coal leasing activity through the 1960s resulted in vast acreages of public land falling under the control of coal land speculators, several western regional Bureau of Land Management offices unilaterally stopped issuing leases. These ad hoc regional moratoria became nationwide in May 1971 when Interior instructed all regional offices to refrain from processing lease applications. On February 17, 1973, then Secretary of Interior Rogers Morton implemented a new leasing moratorium—not a blanket ban on the issuance of leases but a serious limitation of the access of private coal developers to federal land, until a new long-term leasing program could be implemented. His program established "short-term leasing criteria" to permit the issuance of leases for land adjacent to existing mining operations where the mining company needed the additional coal to fulfill current or projected contract requirements. New leases could also be issued to any coal operator who could begin mining operations on the leased land within three years.

The long term leasing strategy that Morton's program anticipated is still being debated—now by the third Administration to study the problem. Interior first outlined new guidelines for a long term leasing program in May 1974 in its Draft Environmental Impact Statement (EIS) on the Proposed Federal Coal Leasing Program. Under the proposed Energy Minerals Allocation Resource System (EMARS), the Department was to project national coal needs and determine the amount of federal coal required to fuel this demand; to allocate the required federal coal reserves among the western regions, taking into account the environmental, social and economic desirability of mining in each region; and then to choose precise public land tracts capable of fulfilling a quota for leasing.

Beset by ambiguities and vagaries, this EMARS program was roundly criticized in the 117 written comments and 37 oral testimonies submitted to the Department of Interior. Many critics—including CEP—complained that Interior had not demonstrated a national need for additional coal leasing in view of the billions of tons of coal leased in the 1960s. Others noted that Interior lacked the regulatory tools needed to keep a new leasing program responsive to national energy interests, as opposed to the economic interests of the western coal lessees.

Ignoring most of these criticisms, the Department of Interior issued a Final Environmental Impact Statement on September 19, 1975 that signalled a retreat from its attempt to manage coal resources. This EIS abolished the one aspect of the old EMARS program which environmental and citizens groups had praised: the selection by government rather than industry of lease sites. EMARS II relieved Interior of the responsibility for determining national coal needs and allocating leases among western coal regions in accordance with government-established production goals. Instead, Interior would rely on coal industry representatives to nominate land tracts for leasing. Interior would then evaluate these requests against "disnominations" submitted by the general public in determining the number and location of leases to be issued.

Western land owners and environmentalists were outraged by the proposed EMARS II program. The Natural Resources Defense Council (NRDC) complained that EMARS II was still too vague, and that alternatives to the leasing of energy resources were not adequately discussed. On October 21, 1975, NRDC filed suit to enjoin the implementation of the EMARS II program until another environmental impact statement had been completed. *NRDC v. Hughes* dragged on for two years. On September 27, 1977, U.S. District Court Judge John Pratt granted a summary motion in favor of NRDC, ordering Interior to prepare another Environmental Impact Statement.

Interior responded by reissuing the old, rejected impact statement for public review. Interior is now rewriting this statement. The new draft will also be circulated for public review and commentary. If the new statement is approved, Interior will officially begin to implement a new coal management policy culminating in the first lease sales—possibly in 1980.

SHORT-TERM CRITERIA

While Interior has been hammering out a new overall leasing strategy, the short-term leasing criteria have been adjusted several times. The short-term criteria first established by then Secretary Morton limited coal leases to companies which could establish mining operations within three years or which needed coal land adjacent to existing mining operations to fulfill current or projected contract needs. In January 1976, then Secretary of Interior Thomas Kleppe attempted to terminate Morton's short-term criteria and announced that new competitive and preference right leases would be issued under EMARS II provisions. Kleppe proceeded to authorize the issuance of several new competitive and preference right leases. NRDC claimed these issuances were illegal, since its lawsuit challenging EMARS II was still pending.

Morton's criteria remained in effect until July 25, 1977. At that point, Interior, under heavy pressure from NRDC and Judge Pratt, issued revised short-term leasing criteria much stricter than the program Morton had established. The new criteria permitted leasing only where the operator of an existing mining operation could demonstrate that additional coal was needed to sustain a mine at current production levels—not projected production increases—to meet current contracts. Only an eight-year supply of reserves could be leased at one time. Leases could be issued for new mines only if no new transportation facilities were needed. Judge Pratt further strengthened the leasing criteria in his September 1977 decision by limiting all new leases to those that covered only three years of coal supply.

NRDC and the Department of Interior agreed on a set of short-term leasing criteria slightly looser than those mandated by the Pratt verdict, and petitioned Judge Pratt to replace his system with theirs. Under the Interior-NRDC criteria for proposed leasing, the Department of Interior would be permitted to issue seven specifically-designated "hardship leases" to provide additional reserves for existing coal mining operations. These seven leases could provide a combined output of 11 million tons per year. A second criterion would allow the Department to lease small isolated land tracts surrounding existing mines if diverting the mining operations around these acreages would have significantly higher economic or environmental costs than mining through the tract. Only five years of reserves could be included in these "bypass leases." Even so, the government expects to lease 40-60 million tons of coal reserves in bypass leases. Third, the Interior Department could issue three-year "employment/contract leases" for land adjacent to existing mining operations, if the lease is needed to keep a mine in operation or to fulfill existing contracts. Finally, the Interior Department would be permitted to process 20 of the 180 currently-pending preference right lease applications based on prospecting permits issued before the 1971 leasing moratorium. Judge Pratt agreed with these suggestions and the new short-term leasing guidelines took effect in July 1978.

FEDERAL LEASING PROGRAM IN ACTION

With a multiplicity of lawsuits, environmental impact statements and increasingly restrictive short-term leasing criteria commanding Interior's attention over the last five years, it would be easy to accept at face value the claims of many mining companies that the leasing "moratorium" has placed a lid on western coal development. CEP, however, has found otherwise—and decidedly so. Several new federal leases have been issued, many existing leases have been sold to new coal development interests, dozens of new mines on leased land have been announced, and western coal output has skyrocketed.

Extent

CEP's survey of the federal leasing program in the six western states found 468 valid federal coal leases covering 703,193 acres of federal land at the end of 1977. This represents a net increase of five over the number of leases and 3% over leased acreage in 1973 as reported in *Leased and Lost* (see Tables 1-3 and 1-4).

Since 1973, 13 new leases spanning more than 30,000 acres have been created—an increase offset by the relinquishment of eight leases. Most of the relinquished leases were among the oldest valid leases, had long been inactive, and were owned by local individuals and their families (see Table 1-5).

Utah and Wyoming, each with over 200,000 acres of federal land under lease, account for almost 70% of all federal land under lease in the West. North Dakota and Montana are the states with the fewest acres under lease, with slightly more than 50,000 leased acres combined. Total acreage under lease has declined slightly in Colorado and by a substantial 5% in North Dakota.

TABLE 1-3

FEDERAL COAL LEASES

	Number of Leases	Change Since 1973
Colorado	113	+ 1
Montana	17	0
New Mexico	28	0
North Dakota	16	− 4
Utah	198	+ 3
Wyoming	96	+ 5
TOTAL	468	+ 5

FEDERAL LAND UNDER LEASE

	Leased Acreage	% Change Since 1973	% of State's Federal Coal Land Under Lease
Colorado	121,417	− 0.4	1
Montana	36,293	0	≪ 1
New Mexico	40,955	0	1
North Dakota	15,432	− 5	< 1
Utah	274,275	+ 3	7
Wyoming	214,821	+ 7	2
TOTAL	703,193	+ 3	1.2

TABLE 1-5

NEW FEDERAL LEASES AND RELINQUISHMENTS
1973-1977

	New Competitive	New Preference Right	New Segregated	Relinquishments
Colorado	1	1	0	1
Montana	0	0	0	0
New Mexico	0	0	0	0
North Dakota	0	0	1	5
Utah	1	1	3	2
Wyoming	0	1	4	0
TOTAL	2	3	8	8

Most federal coal reserves remain unleased. The Department of Interior estimates that aproximately 1.2% of all coal-bearing federal land in the six states surveyed is now under lease. Utah is the only state where more than 2% of federal coal acreage has been leased—7% is currently covered by valid government contracts.

Issuance Procedures

Throughout the leasing moratorium, the Department of Interior has approved the issuance of leases utilizing three procedures in its short term program.

Thirteen leases spanning 30,780 acres of public coal land have been transferred to private individuals and corporations during the so-called leasing "moratorium" (see Table 1-6).

Preference Right Leasing

Preference right leasing, an antiquated, non-competitive land give-away program authorized by the Mineral Leasing Act of 1920 to foster exploration in the West, has accounted for nearly 94% of all the new acreage to come under federal lease since 1973. Under this program, the Department of Interior can issue a permit allowing a corporation or individual the exclusive right to prospect for coal on any land whose mineral content has not been evaluated by the federal government. The permit is valid for two years, and there is an option to lease the land if coal is discovered. The total fee for obtaining a prospecting permit is $10.

In 1973 CEP found preference right leasing to be a very popular mechanism for developers to obtain public coal land. Forty-five percent of the land under lease at that time had been leased by the preference right method. In the two years immediately prior to the 1971 moratorium, 180 prospecting permits covering fully 496,000 acres and 12 billion tons of public coal reserves were issued. Many of these prospectors subsequently requested preference right leases based on the discovery of coal on this land. Interior has deferred action on nearly all of these requests, leaving them to swell into one of the Department's largest headaches. The resolution of these preference right lease applications will have broad implications for any future leasing program. Granting all pending lease applications is clearly an inadequate alternative. Such a

TABLE 1-6

FEDERAL LEASES ISSUED SINCE 1973

Company	Location	Acreage	Date of Issuance	Issuance Procedure
Antelope Coal Co.	Campbell, WY	320	8-12-77	segregated
Black Butte Coal Co.	Sweetwater, WY	14,902	4-1-76	preference right
Carter Oil Co.	Campbell, WY	80	2-10-75	segregated
Consolidation Coal Co.	Campbell, WY	840	2-1-77	segregated
Consolidation Coal Co.	Mercer, ND	322	9-1-77	segregated
El Paso Natural Gas Co.	Kane, UT	640	3-19-74	segregated
Energy Fuels Corp.	Routt, CO	474	6-1-75	preference right
Franklin Real Estate	Carbon, UT	634	5-3-74	segregated
Franklin Real Estate	Carbon, UT	543	7-1-74	segregated
Medicine Bow Coal Co.	Carbon, WY	1,280	12-11-73	segregated
Mountain States Resource Corp.	Sevier, UT	8,824	5-1-77	preference right
Plateau Mining Co.	Emergy, UT	1,360	5-1-74	competitive
Western Slope Carbon Inc.	Gunnison, CO.	241	12-1-73	competitive

policy would increase the amount of public coal under lease by nearly 60% while bringing the government a total payment of less than $2,000.

Although painfully aware of its history of land mismanagement, Interior nevertheless has granted three new preference right leases since 1973. The first lease, to Energy Fuels Corp., covered a small tract in Routt County, Colorado. The company claimed to need the 474-acre tract to continue an ongoing mining operation in accordance with Interior's short-term leasing criteria. Energy Fuels mined over one million tons of coal from the land in 1976 and 1977. Production has now ceased on the lease.

The Department of Interior demanded no such compliance with the short-term leasing criteria before issuing the other two new preference right leases. Both leases were issued by then Secretary of Interior Kleppe in the hiatus between his attempted cancellation of the short-term criteria and Judge Pratt's establishment of stricter leasing criteria. Both lease tracts are huge; both leases were issued non-competitively. One, covering 14,902 acres in Sweetwater County, Wyoming, was issued to Rosebud Coal Sales Co. on April 1, 1976. The other—leased to Mountain States Resource Corporation—covers 8,824 acres in Sevier County, Utah. It was issued on May 1, 1977. These leases rank as the second and sixth largest leases ever issued by the U.S. Government in the history of the federal leasing program. They represent capitulations to the pressure of special interest groups and a return to the ''business as usual'' leasing historically practiced by the Department of Interior.

Events surrounding the issuance of the Rosebud Coal Sales Co. lease are particularly enlightening. In a press release at the time, Kleppe declared confidence that Rosebud would begin mining operations on its land shortly, that contracts to supply coal to utility companies in neighboring states had been formalized, and that issuance of the lease was necessary to fulfill those contracts. This was not the case, however. The contracts with the utilities were in fact merely letters of intent, and most evaporated after the lease was signed. Rosebud does not even own the lease anymore. In 1976 it sold the rights to the lease to Black Butte Coal. Black Butte is now preparing a mining plan for the land and is looking for long term contracts with utilities as far away as Chicago.

Competition

More than half of all outstanding coal leases have been issued competitively—although CEP has found most competitive lease sales to be competitive in name only. Competitive leasing procedures direct the Bureau of Land Management (BLM) to advertise the availability of certain tracts of land for competitive leasing. Although the Department has the authority to select these land tracts, it has never done so. Instead, Interior has accepted and responded to industry nominations for land tracts for coal leasing. At industry request, Interior sponsors a lease sale. A lease is awarded to any individual or company that offers the highest one-time cash bonus payment at the sale. Rental and royalty rates and other lease terms are established by the BLM and are not subject to bidding.

Until the lease moratorium, competitive leasing procedures were not in fact conducive to competition. CEP found that no formal bidders appeared at 10% of all lease sales through 1973, thereby permitting Interior to issue leases to the original applicant who hadn't offered any cash bonus. In a startling 59% of all competitive lease sales, CEP found that only one bidder appeared. At lease sales attracting zero or one formal bidder, the ultimate lessee paid a paltry $3.31 an acre average in bonus bids for the leases. At only 66 competitive lease sales did more than two bidders appear.

Throughout the 1970s, Interior has fared little better in its efforts to arouse competitive interest in its coal lands. The Department still holds lease sales in response to industry requests. Interior has, however, been more successful in garnering larger cash bonus payments at its most recent competitive lease sales. But these higher bonus payments are less a reflection of actual competition among bidders than of applicants' concern that they not be outbid at the last moment by an unexpected party.

Western Slope Carbon, Inc., for example, bid $70 per acre for the 241-acre tract it acquired in late 1973, even though it turned out to be the only bidder at the sale. The company now operates the Hawk's Nest mine on this land and has produced over five million tons of coal to date. Plateau Mining Co., a United Nuclear subsidiary, bid $257.36 per acre for the 1,360-acre tract it leased in May 1974. Only one other bidder appeared at this sale. Plateau has yet to begin mining operations on this land.

Interior held another lease sale in February 1978 in response to Westmoreland Resource's interest in acquiring a 310-acre tract adjacent to its Orchard Valley mine in Colorado. Westmoreland claimed to need the coal to meet contracts it had signed with Northern Indiana Public Service Co. Of-

TABLE 1-7

FEDERAL COMPETITIVE LEASING

Number of Bidders	Number of Leases	Acreage Leased	Average Bid per Acre
0	24	14,885	$ 0.00
1	141	181,470	5.39
2	38	69,261	29.82
3	13	42,363	32.96
4 or more	14	45,012	179.27
NA	9		
TOTAL	239	352,991	$ 35.44

fering $100.50 per acre, Westmoreland was the successful bidder for the tract at the lease sale. (This lease was issued too late to be included in CEP's Lease Catalogue). The only other offer was a token protest bid submitted by the Rotten Apple Coal Co. Mining on the tract has already begun.

From the beginning of the coal leasing program in 1920 through the end of 1977, the Department of Interior has issued 239 competitive leases covering 352,991 acres. This land has been leased for an average of $35.44 per acre. The size of the winning bid is directly proportional to the number of bidders appearing at the lease sale. An average bid of only $5.39 per acre was obtained at each of these 141 competitive lease sales at which only one bidder appeared (see Table 1-7). On the other hand, a much greater $179.27 per acre bid was received on the average at the 14 lease sales at which four or more bidders appeared.

Segregation

Segregation is the process by which an existing lease is split into two new leases, the acreage total of both equalling the acreage of the original lease prior to segregation. The original lessee can retain ownership of one, both or none of the resultant leases. If the original lessee retains some ownership, the remnant of the old lease is termed a "modified" lease while the new lease is the "segregated" lease. The Department of Interior receives a segregation fee of $10 to defray the paperwork involved in issuing the new lease contracts.

Eight new leases have been created by segregation since 1973. Consolidation Coal has acquired two of these, both issued in 1977: one in Campbell County, Wyoming and the other in Mercer County, North Dakota. American Electric Power's subsidiary Franklin Real Estate has also acquired two segregated leases, both created in 1974 and located in Carbon County, Utah. The new leases were segregated from holdings of Bratzah Corp. and Spring Canyon Coal Co. and were subsequently transferred to Franklin Real Estate. In 1977 Antelope Coal acquired a 320-acre tract segregated from a Peabody Coal Co. lease in Wyoming. Carter Oil Co., a subsidiary of Exxon, received ownership of an 80-acre tract formerly part of a Meadowlark Farms-owned lease (Meadowlark Farms is a subsidiary of AMAX). El Paso Natural Gas Co. was assigned ownership of a segregated lease created in 1974 out of a lease owned jointly by Mono Power, Resources Co., and New Albion Resources Co.—all subsidiaries of Arizona Public Service and San Diego Gas & Electric. Finally, Ark Land segregated part of one of its leases in Carbon County, Wyoming late in 1973, transferring ownership of 1,280 acres to Medicine Bow Coal Co. This acreage is now part of the Medicine Bow mine.

Assignment

There is another, better-travelled route by which interested individuals and companies have been able to obtain federal coal leases throughout the moratorium. Under the terms of the Mineral Leasing Act, lessees are permitted to sell their leases at any time, to any party. These lease "assignments" are accompanied by cash transactions between assignors and the assignees. Many assignments also include an overriding royalty clause requiring the new lessee to pay royalties to the original lease owner on top of those paid to the federal government. The government, for its part, receives no cash or any other consideration from the assignment transaction. Indeed, Interior exerts no authority over the assignment process.

No single provision of the Mineral Leasing Act has done more to foster the speculation of public coal land than the assignment provision. It invites into the coal leasing program, companies seeking to profit from their public coal holdings not by developing them, but by locking them up with only minimal investments—rental rates seldom exceed $1 per acre per year—and then selling them when market changes enhance their value. The legality of the assignment provision, together with Interior's long-established willingness to indiscriminately lease to any party expressing an interest in public coal land, the low costs of obtaining and maintaining leases, and the absence of diligent production requirements have acted to encourage speculation and discourage production on federal coal land. CEP discovered in 1973, for example, that only 10% of all outstanding coal leases were producing coal, while 321 of the 463 then-valid leases had never produced even a single ton.

CEP found that in the last five years an incredible 119 of the 463 federal coal leases have been bartered on the assignment market. During the moratorium, more than one-fourth of all valid coal leases changed hands, allowing several new companies to make the list of top federal leaseholders. For example, Franklin Real Estate, an American Electric Power subsidiary, has picked up all 11 of its leases by assignment since 1973. So while American Electric Power has been among the most vocal critics of the leasing moratorium, the utility holding company has actually prospered well under it, obtaining rights to 12,159 acres of public coal land by assignment—more than any other company.

CEP found further that lease assignments conducted over this period tended to result in leases moving from small companies and individuals to major energy conglomerates. Sun Oil Co.'s subsidiary, Sunoco Energy Development Co., the tenth largest leaseholder in 1973, has acquired five additional coal leases via assignment during the moratorium. Three were purchased from Humac Corp. and two from Summit Exploration and Development Co. AMETEX, a national diversified conglomerate, has acquired four leases in New Mexico by assignment, two from Florentino Padillo, one from John Sacket, and one from the Elena Mining Corp. Utah Power & Light has picked five leases off the assignment market. Coal Search Corp. has acquired four—all from Jessie Knight. Getty Oil bought a lease from Marjorie Mann. Gulf bought out W. B. Brannan's lease, as well as six leases owned jointly by J. C. Karcher and Concho Petroleum. And the list goes on and on. A complete inventory of lease assignments is included in CEP's Lease Catalogue.

TABLE 1-8

FEDERAL LEASE ASSIGNMENTS SINCE 1973

	Number of Assignments	% of Leases
Colorado	20	18
Montana	2	12
New Mexico	17	61
North Dakota	3	18
Utah	40	20
Wyoming	37	39
TOTAL	119	25

Assignments have occurred in every western state during the moratorium. Of the 119 leases assigned in the last five years, Utah with 40 and Wyoming with 37 together account for fully 65%. In New Mexico, where no new federal leases have been issued in the past five years, a whopping 61% of all valid leases have changed hands (see Table 1-8).

Coal Mining Boom

While the speculative holdings of federal leases is still rampant, CEP has found that there has been a massive upsurge of coal development on public land in the last five years. Markets for western coal among midwestern and far western utilities are increasing at a rapid rate, and there has been a boom in construction of coal-fired power plants in the coal producing states themselves. The inadequacies and maladministration of the federal leasing program to date have prevented the federal government from satisfactorily regulating this development. Millions of dollars in potential royalties and cash bonus payments have been lost to federal treasuries by the leasing of so much federal land prior to the development of a competitive market for western coal.

There has been a 41% increase in the number of leases sustaining mining operations from 1973 to 1977 (see Table 1-9). Nevertheless, these producing leases constitute just 14% of all outstanding leases. That 86% of all federal leases are still inactive, even in the midst of a coal mining boom, indicates the vast reserves of leased but unmined federal coal.

In every western state except North Dakota, more leases produced coal in 1977 than in 1973. Mining activity on federal land in Wyoming and New Mexico doubled in this period. More than one-third of the outstanding leases in Montana are now producing coal, making it the western state with the greatest percentage of active leases. Utah has more active leases—21—than any other state, although its percentage of active leases is the lowest in the West.

Even more dramatic than the increase in the number of active leases has been the increase in coal production on federal land. Total output in 1977

TABLE 1-9

FEDERAL LEASES PRODUCING COAL IN 1977

	Number of Active Leases	% of All Leases	% Change Since 1973
Colorado	14	12	+27
Montana	6	35	+50
New Mexico	4	14	+100
North Dakota	4	25	-50
Utah	21	11	+50
Wyoming	18	19	+100
TOTAL	67	14	+41

TABLE 1-10

COAL PRODUCTION ON FEDERAL LAND

	1977 Production (tons)	% Change Since 1973
Colorado	3,890,829	+21
Montana	10,459,537	+370
New Mexico	2,849,042	+472
North Dakota	740,464	-32
Utah	5,818,033	+96
Wyoming	28,685,688	+434
TOTAL	52,443,593	+241

from the 67 producing leases was 52,443,593 tons, representing a 241% jump in coal output over the 15.3 million tons produced in 1973. Furthermore, more than twice as much coal is being mined from the average producing lease now—782,000 tons per lease—than in 1973 when the average producing lease yielded about 333,000 tons of coal (see Table 1-10). The opening of several massive strip mining operations, primarily on public land, in the mid-1970s helps explain these increases. Four of these mines—Belle Ayre, Decker, Colstrip and Jim Bridger—though only a few years old, now rank among the six largest coal mines in the country. Indeed, the three largest coal mines in the country are now located at least partially on federal land.

The 10 largest mining operations generated 76%, or 39.7 million tons, of the 52.4 million produced on federal land in 1977. All 10 are strip mining operations. Several mines span more than one lease.

Five of the 10 largest mines operating on public land in the West in 1977 are located in Wyoming, where the coal boom is at its most intense (see Table

1-11). These five mines contributed almost half the output from all federal land in the six western states, while accounting for 62% of the output of the 10 largest federal mines.

AMAX's Bell Ayre mine in Campbell county, Wyoming, produced 13.4 million tons of federal coal in 1977—far more than any other mine in the West. Although production began there only in 1973 with just 219,000 tons, the Belle Ayre mine now ranks as the largest coal mine in the country. Its 1977 output represents more than one-fourth of all western public coal output for the year and more than twice the output of the largest mine in the country a scant five years ago.

The Jim Bridger mine, operated by the Bridger Coal Co. in Sweetwater County, Wyoming, feeds coal into the neighboring Jim Bridger power plant; it added 4.4 million tons to federal coal production last year. Inactive in 1973, the Bridger mine began production the following year. It now ranks as the sixth largest coal mine in the country. Somewhat smaller is Pacific Power & Light's Dave Johnston mine—the fifth largest mine operating on federal land in 1977—which produced 3.2 million tons. Coal mined there, in Sweetwater County, Wyoming, supplies the nearby Dave Johnston power plant. The mine has expanded from one federally-leased tract in 1973 to a second in 1974 and a third in 1977. All three tracts contributed to the mine's output in 1977. Sunoco Energy Development Co.'s Cordero mine and Ark Land's Seminoe II mine, both also in Wyoming, rank ninth and tenth among the largest mines on federal land in 1977. Seminoe II mining operations have spread onto a second

TABLE 1-11

TEN LARGEST MINES ON FEDERAL LAND IN 1977

Company	Mine	Location	1977 Federal Output (tons)
AMAX	Belle Ayre	Campbell, WY	13,389,615
Bridger Coal Co.	Jim Bridger	Sweetwater, WY	4,425,291
Decker Coal Co.	Decker	Big Horn, MT	4,307,431
Western Energy	Colstrip	Rosebud, MT	3,960,221
Pacific Power & Light	Dave Johnston	Converse, WY	3,236,616
Energy Fuels Co.	Morgan Strip	Routt, CO	2,467,474
Western Coal Co.	San Juan	San Juan, NM	2,246,041
Peabody Coal Co.	Big Sky	Rosebud, MT	2,144,983
Sunoco Energy Dev. Co.	Cordero	Campbell, WY	2,128,543
Ark Land	Seminoe 2	Carbon, WY	1,453,472
		TOTAL	39,759,642

federal lease since 1973, while production just began on the Cordero tract in 1976.

Three of the 10 largest mines on federal land are located in Montana, making that Northern Plains state the second largest contributor of federal coal in the West. The portion of the Decker mine located on federal land in Big Horn, Montana, produced 4.3 million tons in 1977. Western Energy, a subsidiary of Montana Power Co., operates the Colstrip mine in Rosebud, Montana, where 1977 federal production totalled 3.9 million tons. Federal output on the Decker and Colstrip mines helped make them the second and third largest coal mines in the country in 1977. Peabody Coal Co.'s Big Sky mine generated an additional 2.1 million tons of federal coal in Montana.

Also among the top 10 federal mines in 1977 are Energy Fuels Co.'s Morgan strip mine in Colorado and the San Juan mine of Public Service of New Mexico's subsidiary, Western Coal Co. Both mines have expanded production since 1973 by stretching onto a second federally-leased tract.

North Dakota's mere 740,000 ton federal output from its four active leases represents the lowest production level of the six states surveyed—and makes it the only western state to register an output decline since 1973. With the fewest number of outstanding leases covering the smallest acreage under lease yielding the least amount of coal—and with the number of leases, acreage under lease, and federal coal output all in steady decline—North Dakota stands in stark contrast to the other western states, all undergoing coal booms of varying intensities. CEP believes that the decline of interest in North Dakota's federal coal represents a market adjustment to current expectations of western coal developers. Virtually all coal found in North Dakota is lowgrade lignite which is generally uneconomical to transport the long distances necessary to feed midwestern or far western utilities. There are, moreover, limited markets for, and intense local opposition to, the burning of coal within the state at large power plants.

Perhaps more important, much of the early interest in North Dakota's lignite deposits stemmed from its suitability for coal gasification. High moisture content and non-caking qualities allow lignite to perform more favorably in gasification units than any other grade coal. But in the past five years gasification has been criticized as an expensive, environmentally unsound technology. Most of the large coal conversion schemes announced in the early 1970s have been either delayed or cancelled.

Even if North Dakota is being left behind, the federal coal fields in the West are now producing more coal than ever before. This upsurge in the last five years reflects the growing importance of federal coal in the western and national coal mining industries. As late as 1973, federal coal production in the six western states accounted for 28% of western output and 2% of the nation's coal output from mines on private, state, Indian and federal lands combined. By 1977, federal production in the west contributed 46% of the western and 8% of the national coal output.

Revenue

The first source of revenue accruing to the Department of Interior from coal leasing activity is from the lease issuance procedure. From 1973 to 1977, filing fees for the creation of the eight new segregated leases brought the Department $80. The issuance of the three new preference right leases brought in an additional $30. Cash bonus payments from the issuance of the two new competitive leases were $366,879. Thus, total income associated with the issuance of leases since 1973 amounted to $366, 989, or just $12.05 for every acre added to the federal leasing program over this period.

Annual rental payments from the lessees to Interior constitute the second source of lease revenues. Holders of inactive leases are required to pay rent on their leased land. Several rental structures are in effect, but most lessees are required to pay $1 per acre per year in rent. CEP estimates total rental payments collected in 1977 from the 468 valid coal leases to be less than $700,000.

Lessees producing coal from public land are required to pay a royalty to the Department of Interior based on every ton of coal extracted. In 1977 the Department collected over $9.8 million in royalties (see Table 1-12). The average royalty rate was 18.8¢ per ton. With the selling price of western coal hovering around $15 per ton, this royalty rate represents only 1.3% of the value of the coal. Still, this is a 50% improvement over the 12.5¢ per ton collected by the federal government through 1972.

In the early 1970s, the Department of Interior switched from a fixed to a variable royalty schedule. The fixed royalty system required holders of producing leases to pay a specific amount for each ton of coal extracted—10¢ or 15¢, for example—regardless of the selling price of the coal. The variable royalty system is based on a percentage of the coal's selling price. The first leases Interior issued with variable royalty provisions required payment of 5% of the gross value of the coal. Leases issued in the early 1970s included an 8% royalty provision. The Mineral Leasing Act Amendments of 1975 require the Department of Interior to collect at least 12.5% of the value of the coal mined from new leases.

TABLE 1-12

ROYALTIES FROM FEDERAL LAND COAL PRODUCTION

	1977 Royalties	Royalty per Ton
Colorado	$ 706,571	18.1¢
Montana	1,627,299	15.6
New Mexico	648,042	22.7
North Dakota	157,715	21.3
Utah	963,861	16.6
Wyoming	5,767,434	20.1
TOTAL	$9,870,922	18.8¢

Unfortunately for the U.S. Treasury, Departmental regulations concerning royalties are not retroactive: they apply only to new leases and to leases renewed after their existing contracts expire. Federal coal leases are issued for 20-year terms, but generally may be renewed. At the time of renewal, the federal government has the opportunity to alter the terms of the lease contract to include higher royalty rates. But on the 35 leases that have come up for renewal of terms—eight of which are currently producing coal—Interior has failed to adjust the royalty rates. The Department has preferred to extend the existing lease contracts with the old "fixed royalty" terms until the advent of a new government leasing program. If current coal miners operating on leases which authorize a fixed royalty schedule were required to pay 12.5% of the value of their coal instead, the federal government would be collecting approximately $1.00 more in royalties for every ton of coal produced. In 1977 alone, the eight producing leases whose terms could have been adjusted but weren't generated more than 4.1 million tons of coal; had Interior raised their royalty rates, it would have collected an additional $4.1 million. With the higher royalties included only in new leases, federal royalties in 1977 remain, for the most part, embarrassingly low.

Ownership

CEP has found little change in the overall level of concentration of ownership of federal leases since 1973. *Leased and Lost* noted extensive consolidation of lease holdings among a few corporations. While there are now fewer lessees of federal coal land than there were in 1973—an indication of further concentration—control of leased acreage among the top five lease holders has also dropped, suggesting the opposite. In 1973, the top five lessees controlled 37% of all federal coal land under lease. By 1977, their share dipped to 31%, or 218,209 acres (see Table 1-13).

Peabody Coal Co. is the top federal leaseholder, as it was in 1973, controlling 43 leases covering 71,371 acres—more than 50% more land than any other federal leaseholder controls. (Peabody is also the number one holder of leases on Indian land in the West, controlling three of ten outstanding leases.) Peabody was a subsidiary of Kennecott Copper Co. until 1977, when the Justice Department's Antitrust Division ordered Kennecott to divest Peabody. The coal company was then acquired by a consortium of industries including Newmont Mining Corp., Williams Co., Bechtel Corp., Boeing Co., Fluor Corp., and Equitable Life Assurance Society. Under the consortium's control, Peabody remains one of the fastest growing coal companies in the country, with extensive holdings on Indian and public lands west of the Mississippi.

Consolidation Coal Co., a subsidiary of Continental Oil, remains the second largest holder of federal leases. Consol is also the number one leaseholder of state-owned land in Montana and the number three leaseholder of state-owned land in Colorado. It has acquired two new leases since 1973, both segregated in 1977 from existing leases. Consol has transferred five leases

TABLE 1-13

FIVE LARGEST FEDERAL LEASEHOLDERS*

Company	Number of Leases	Acreage under Lease
Peabody Coal Co.	43	71,371
Consolidation Coal Co.	32**	46,675***
Resources Co.		
Mono Power	21	40,276
New Albion Resources Co.		
Kemmerer Coal Co.	26**	32,220***
El Paso Natural Gas Co.	16	27,657
	TOTAL	218,209

*The top five federal leaseholders control 31% of all leased federal acreage.
**Includes leases fully and jointly owned.
***Acreage computed on the basis of percent of ownership of lease.

in Sandoval County, New Mexico to Ideal Basic Industries. Terms of the assignments are unknown. It continues to hold 10 coal leases jointly with Kemmerer Coal Co.—the fourth largest federal coal lessee in its own right.

The three company venture of Resources Co.—a subsidiary of Arizona Public Service—Mono Power and New Albion Resources Co.—both subsidiaries of San Diego Gas & Electric—is still the third largest federal leaseholder. The consortium controls 40,276 acres of federally-leased land in the West. Kemmerer Coal Co., again the fourth largest federal leaseholder in the six states, controls 32,220 acres.

El Paso Natural Gas Co.'s acquisition of a 550-acre lease has allowed it to break into the list of the top five federal leaseholders. Its 16 leases spanning 27,657 acres represent the fifth largest federal lease holdings in the West. Pacific Power & Light, the fifth largest federal leaseholder in 1973, has since dropped from the list of the top five by assigning more than half its holdings. Five of its 17 leases in 1973 have been assigned to Resource Development Co. Another three have been transferred to Bridger Coal Co., which operates a mine in southern Wyoming. A final lease owned by Pacific Power & Light in 1973 has been assigned to Decker Coal Co., which it controls in a joint venture with Peter Kiewit Sons.

While changes among the top five federal leaseholders have been limited, major changes in overall lease ownership have occurred over the past five years. Sun Oil Co. has added 4,000 acres obtained by assignment to the 21,000 acres it controlled in 1973. Gulf Oil Co.'s holdings have jumped from 9,430 acres in 1973 to 16,442 acres in 1977, moving the oil company up from twelfth to tenth largest leaseholder. Gains by Gulf and Sunoco point to the expanded involvement of major oil companies in the western coal development picture.

The top 15 federal leaseholders in 1977 include five oil majors: Continental Oil (through its subsidiary, Consolidation Coal Co.), Sunoco (through Sunoco Energy Development Co.), Gulf, Exxon (through Carter Oil Co.) and Atlantic Richfield.

The number of federal leaseholders has meanwhile dropped from 144 in 1973, to 130 in 1977. The reason for this decline is that several individuals who held only one lease, usually an older, inactive one, have assigned their leases to a few, large corporate interests. W. B. Brannan, Elaine Poulos, and J. C. Karcher all assigned their coal leases to Gulf Oil since 1973, for example, thereby adding eight leases to Gulf's holdings while reducing the number of federal leaseholders by three.

NEW DIRECTIONS FROM WASHINGTON

CEP believes that the federal leasing moratorium has done little to stem the chaotic administration of the program since 1973. Government actions concerning federal coal lands have been random, ad-hoc, unsystematic and uncomprehensive—their only consistency, a failure to reflect any clear animating policy. Meanwhile, a complex web of lawsuits, legislation and governmental regulations has developed in the past five years—all of which will profoundly affect federal leasing policy as it begins to re-emerge.

Litigation

Three major legal actions presented in the District Court in Washington, D.C., directly affect the future of public land coal leasing. *Sierra Club v. Morton* concerned the limits of the Department of Interior's responsibility for the preparation of Environmental Impact Statements not only for individual projects but for whole regions slated for large scale development. The *NRDC v. Hughes* ruling requires the Department of Interior to complete an adequate Environmental Impact Statement on the proposed federal coal leasing program before leasing can resume. *NRDC v. Berkland* seeks to determine the Department of Interior's policy towards the resolution of the 500,000 acres of federal coal land now under preference right lease application.

Sierra Club v. Morton

In June 1973 the Sierra Club and five other environmental groups filed suit in U.S. District Court in Washington, D. C., asking that the Department of Interior be required to prepare an Environmental Impact Statement analyzing the total impact of its resource development program—from coal leasing to reservoir building and other actions—on the Northern Plains. The Sierra Club argued that the impact of coal development in the West would be greater than the sum total of the impact of various individual projects—each of which must have their own Environmental Impact Statement. They cited the National Environmental Policy Act (NEPA) of 1969 which directs government agencies to prepare Environmental Impact Statements prior to any major federal action significantly affecting the environment.

In February 1974 the District Court ruled against the Sierra Club. One year later, however, a three-judge panel of the Circuit Court of Appeals reversed the ruling, noting that "when the federal government, through its power to approve leases, mining plans, rights-of-way and water option contracts, attempts to 'control development' of a definite region, it is engaged in a regional program constituting major federal action within the meaning of NEPA, whether it labels its attempts by plan, a program, or nothing at all." Interior labelled the decision "disastrous," claiming it would delay development "for years." Indeed, the Appeals Court enjoined Interior against the issuance of any new federal lease permits.

Interior petitioned the Supreme Court in late 1975 to review this decision. In January 1976 the Supreme Court lifted the Appeals Court injunction. By a seven to two decision later in the year, the Supreme Court rejected the arguments of the Sierra Club and the Appeals Court, and ruled that no overall impact analysis was required under NEPA guidelines.

The significance of *Sierra Club v. Morton* was that it slowed the pace of western coal development, as the Department was prevented from processing permit and lease applications. The Supreme Court decision, however, cleared the way for the regulatory process to continue—as long as Interior completed Environmental Impact Statements for each specific mining project and a Programmatic Environmental Impact Statement describing the whole coal leasing program.

NRDC v. Hughes

The Natural Resources Defense Council (NRDC) filed suit on September 19, 1975 in response to Interior's publication of its Final Environmental Impact Statement on the new coal leasing program, a study that NRDC and most other reviewers thought grossly inadequate. NRDC charged that the new program was insufficiently explained, and that alternatives to leasing to help meet the objectives of the nation's energy program were not considered.

In September 1977 the U.S. District Court ruled in favor of NRDC and ordered the preparation of a new Programmatic Environmental Impact Statement. The present Interior timetable calls for completion of the EIS in early 1979.

NRDC v. Berkland

On March 5, 1975 the Natural Resources Defense Council filed suit to enjoin the Department of Interior from issuing 180 preference right coal leases based on prospecting permits issued before the 1971 moratorium. Interior claimed a legal obligation to grant leases to permit holders who had discovered commercial quantities of coal, even though the permits were issued without land or resource use planning and without the preparation of Environmental Impact Statements. NRDC countered that Interior has the authority to deny preference right lease applications whenever there are overriding environmental problems associated with the mining of coal covered by the permit. It holds that Interior should prepare separate Environmental Impact Statements for each lease application and decide on a case by case basis whether to issue the lease.

At stake are nearly 500,000 acres of land under which lie buried 12 billion tons of coal—equal to about two-thirds of the land already leased throughout the entire 55-year history of the leasing program (see Table 1-14). Such extensive acreages and reserves suggest the profound effect resolution of this suit will have on the short-term leasing program.

In July 1978 the District Court ruled on the case. The decision was an ambiguous one, however, and both sides have claimed partial victories. On the one hand, the Court ruled that Department of Interior is obligated to issue preference right leases to the waiting applicants. On the other hand, the Court specifically noted that the Department could set extremely rigid provisions in the lease terms, which could greatly limit the activities of coal lessees and make coal development on these lease tracts difficult or impossible. The case could be appealed by either party late in 1978.

Interior meanwhile has taken a series of actions affecting preference right lease applications. The department first ordered all lease applicants to submit geological data supporting their claim of discovery of coal in commercially mineable quantities; the deadline for these submissions was November 1977. Second, in August 1977 the Department's Solicitor's Office recommended the invalidation of any preference right lease application which included land covered at any time by a mining claim issued under the provisions of the Mining Law of 1872, because the granting of an earlier claim on this

TABLE 1-14

PENDING PREFERENCE RIGHT LEASE APPLICATIONS*

	Number of Applications	Acreage	Reserves (billion tons)
Colorado	41	93,501	2.5
Montana	8	26,306	.5
New Mexico	26	75,546	.7
North Dakota	0	0	0
Utah	38	138,000	1.2
Wyoming	65	152,814	6.9
TOTAL	178	486,167	11.8

*There are five applications covering 9,500 acres in Oklahoma. All others pending preference right lease applications are in the six states surveyed.

40

land—even for a mineral other than coal—indicates that exploration of the tract's resources occurred prior to the issuance of the prospecting permit. The Mineral Leasing Act of 1920 authorizes prospecting permits only for land on which no exploration data has been collected. Should Interior accept this recommendation, most preference right lease applications now pending could be invalidated. Third, the Department is now considering awarding preference right lease applicants leasing "credits" to be offered as bids in lieu of cash at lease sales in the future. Applicants would, in effect, exchange the land covered by their permits—which was chosen randomly without land or resource planning—for coal-bearing land chosen according to the criteria of the new leasing program.

Federal Legislation

Mineral Leasing Act Amendments

At the root of the historical inadequacies of the federal leasing program are the archaic provisions of the Mineral Leasing Act of 1920. CEP found in *Leased and Lost* that a combination of vague language, gaping loopholes and embarrassingly low lease requirements all conspired to provide the Department of Interior with few guidelines and little direction. In 1974 Congress began drafting amendments to the Mineral Leasing Act to remedy the situation. CEP testified before both House and Senate subcommittees, explaining the findings of *Leased and Lost* and recommending legislative solutions. A set of amendments was voted into law in 1976. Table 1-15 identifies the major provisions included in the amendments and the major problems, as identified in *Leased and Lost,* each was intended to redress.

The Federal Coal Leasing Amendments Act of 1975 endured a long and tempestuous process of legislative and executive scrutiny. The late Senator Lee Metcalf (Dem.-MT) engineered the passage of the first version of the amendments through the Senate in mid-1974, but the House failed to act on a comparable bill. In the next legislative session, Metcalf, undaunted, reintroduced the bill with slight revisions; it passed in July 1975 by a vote of 84-12. Meanwhile, Representative Patsy Mink (Dem.-HI) introduced into the House a set of amendments which promised even more extensive revisions of the Mineral Leasing Act. After the House overwhelmingly passed the Mink bill, the Senate also opted for the stronger bill in place of the Metcalf draft. The amendments were then sent to President Ford.

At 11:20 p.m. on Saturday, July 3, 1976, 40 minutes before the deadline for action on the bill, President Ford exercised his fifty-first veto and nixed the Federal Coal Leasing Amendments Act. Ford described the bill as "littered with many provisions that would insert so many rigidities, complications, and burdensome regulations . . . that it would inhibit coal production on federal lands." Within one month, the Senate and then the House decisively rebuked Ford, overriding a Presidential veto for the tenth time in two years.

41

TABLE 1-15

SUMMARY OF THE FEDERAL COAL LEASING
AMENDMENTS ACT OF 1975

Major Issue Addressed by CEP	New Provision
Speculative holding of land	1. There is mandatory termination of leases not producing coal 10 years after issuance.
Concentration of leases among a few large corporations	2. One corporation can acquire a nation-wide maximum of 100,000 acres of leased land.
	3. A more precise definition of "corporation" prevents the leasing of additional land by holding companies.
	4. 50% of all new leases must be leased by the deferred bonus bidding system. Deferred bonus bidding allows a company to pay for a lease in a series of installments. This allows more smaller, undercapitalized firms to participate in the "front end" capital intensive leasing program.
	5. The Department of Justice is to review all proposed leases prior to their issuance for anti-trust implications. Justice can, in some cases, veto a lease or require additional provisions.
	6. A "reasonable" number of leases are to be reserved for bidding by public entities such as electric "cooperatives."
The ability to obtain fair market value for leased land	7. The preference right leasing system, which allows a company to obtain a lease for $15 if it explores unsurveyed land and discovers coal in commercially exploitable quantities, is abolished. A company is still able to obtain exploration permits, but if it discovers coal, the land will be leased via competitive bidding.
The failure of lease terms to reflect market conditions and government regulations	8. Leases must include a minimum royalty of 12.5% of the current market value of the coal mined from leased land. The current minimum is 5¢ per ton.
	9. Lease terms must be updated every 10 years, as opposed to the current 20 year adjustment periods.
The lack of land and resource use planning prior to leasing	10. No leases can be issued for land which has not been studied as part of a comprehensive land use plan. The EMARS program will provide the mechanism to enforce this provision.
	11. Public hearings must be held in areas where leasing is to occur to consider the advisability of leasing and ways to minimize the impact of coal mining on leased land.
	12. If a state's governor objects to the issuance of a particular lease, Interior must consider the objections before issuing the lease. Should it decide to issue the lease anyway, special leasing procedures must be followed.
The inability to adequately regulate coal development	13. The Department of Interior may consolidate two or more federal leases and adjacent state and private lands into a single "logical mining unit," the development of which would be regulated *in toto* by the Department of Interior.
Devastating social and economic impact on nearby areas from coal development on leased land	14. The percentage of royalties which is dispersed to state governments is increased from 37 1/2% to 50%. The extra 12 1/2% must be earmarked for planning and building public facilities in the affected areas.
Haphazard, uncoordinated leasing	15. The Department of Interior is required to accelerate coal exploration on federal land and to prepare yearly summaries of the leasing program.
	16. The Office of Technology Assessment is instructed to undertake an analysis of the federal leasing system in general.

The passage of the nation's first federal strip mine regulation law in August 1977 marked the end of one of the most ferocious environmental battles of the 1970s. Five years of high-pitched debate over regulation of strip mining and reclamation of strip mined land produced two bills, both passed by Congress only to be vetoed by President Ford. Both vetoes were sustained. Finally, the Carter Administration championed a third attempt by Congress to pass this important legislation. President Carter signed the bill into law in the White House Rose Garden in July 1977. The Departments of Interior and Energy are now engaged in establishing a bureaucracy and promulgating regulations to execute the mandate of the law. These regulations will replace all existing environmental standards on coal mining activities on federally-leased land, unless the existing state regulations prove stricter than the new federal ones.

Section 714 of this new Surface Mining Act, which concerns the issuance of federal leases on land owned privately at the surface, will have a special impact on the leasing program. The Homesteading Act of 1916 and other federal legislation allowed early western settlers to acquire ownership of the surface rights to land in the public domain either for free or for a nominal fee. The federal government often retained ownership of the minerals underlying this land and the right to lease them to other private interests. Approximately 39% of all federal land now under lease is privately owned at the surface.

The legal ramifications have never been resolved in the West, but many experts agree that mining companies and other holders of sub-surface mineral leases are allowed to interfere with surface activities, as long as the surface owner is compensated for any damages incurred by sub-surface activities. There has been widespread fear in the West that mine operators would force ranchers and farmers off their land after payment of only minimal compensation. Section 714 will provide some protection for these surface owners by requiring their consent prior to Interior's issuance of the lease.

Nevertheless, in July 1978 Secretary of Interior Cecil Andrus reversed his earlier position that future leasing be confined to areas where no surface-subsurface conflicts exist. He instructed the Department to institute a special ranking system to give preference to the issuance of leases on tracts where no such conflicts exist. Thus, potential lease tracts privately owned at the surface will still be considered for leasing.

Section 714's impact will be greatest in the Northern Plains (see Table 1-16). Virtually all federal land now under lease in Montana and North Dakota is privately owned at the surface. At the other extreme, only 5% of Utah's leased federal land is subject to surface-subsurface ownership disputes.

"Organic Act"

The Federal Land Policy and Management Act of 1976, the "Organic Act," mainly concerns the federal government's management of national forest and grazing land. It also authorizes the Secretary of Interior to encourage and ap-

TABLE 1-16

**SPLIT SURFACE AND SUBSURFACE OWNERSHIP
ON LEASED FEDERAL COAL LAND**

	Split Ownership Land (million acres)	% of Coal Land Leased
Colorado	54,606	45
Montana	34,967	97
New Mexico	26,198	64
North Dakota	16,436	100
Utah	13,335	5
Wyoming	117,196	59
TOTAL	262,738	39

prove exchanges of public land tracts, including leased coal lands. This provision, as specified in Section 206, permits lease holders and applicants to exchange their randomly-chosen lands—often in areas where mining would be environmentally devastating—for new lease tracts chosen after a comprehensive land use study suggests their environmental attractiveness for mining. Lessees may still choose, however, to stand firm behind their existing leases and refuse to swap them.

Divestiture Legislation

CEP has found that the major oil companies continue to figure prominently among the largest holders of federal coal leases. Congress is considering several pieces of legislation to limit the acquisition of coal reserves by these oil companies. "Horizontal divestiture" bills have been sponsored by Senator Edward Kennedy (Dem.-MA.) and Representative Morris Udall (Dem.-AZ).

The original Kennedy bill proposed to bar the major oil companies from expanding further into coal and uranium markets and ordered them to sell the coal and uranium assets they already possessed. After that bill failed to clear the Senate, Kennedy reintroduced a divestiture provision as an amendment to President Carter's energy package—this time without the order that oil companies divest their existing assets. Nevertheless, it was roundly rebuked, 62-30. Kennedy is still pushing the bill in the Senate Antitrust and Monopoly Subcommittee.

Representative Udall's divestiture bill is less broad in scope than Kennedy's, but it specifically prohibits the further acquisition of coal leases by the major oil companies. Although President Carter favored the bill last year, support subsequently eroded, and he has been silent on the subject lately. The Justice Department Antitrust Division has come out against the bill, although it recommends the invalidation of any lease issued to an oil or gas company or electric utility with more than 15% control of a region's coal market. The Udall bill failed to clear the House Interior Committee in April 1978 by a one vote margin.

Interior Regulations

While the chambers of the nation's highest courts and the halls of Congress have been abuzz with activity concerning coal leasing, the Department of Interior has been working behind closed doors in its effort to contribute to the revamping of the federal coal leasing program. The Department has already prepared several drafts of the Programmatic Environmental Impact Statement it hopes to release in early 1979. It has begun to weigh varying strategies for the dispensation of the 180 preference right lease applications now confronting it. And it is busy clarifying the terms of the Mineral Leasing Act Amendments, initiating individual Environmental Impact Statements, and reshaping the coal leasing bureaucracy.

Since December 1974, Interior has sought to codify and clarify the ambiguous terms Congress was wrestling with in its consideration of the Mineral Leasing Act of 1920 and its 1975 amendments. For example, The Mineral Leasing Act of 1920 directed that coal leases be contingent upon the "diligent production and continuous operation of a mine," but CEP found in 1973 that Interior permitted leaseholders to simply pay one year's advance rent in lieu of production. Interior therefore attempted to comprehensively define "diligent production" and "continuous operation" to insure that coal development could not be avoided through a legal loophole. Hence, diligent production now means "the preparation to extract coal from a logical mining unit," which itself is explicitly defined as a "compact area of coal land that can be developed and mined." Similarly, continuous operation is the "extraction, processing, and marketing of coal in commercial quantities for more than six months in any calendar year. Having established viable definitions, Interior then gave leaseholders two years to organize their lease acreage into logical mining units, and then to report efforts to fulfill diligent production requirements within 30 days after each two-year anniversary of the lease.

After passage of the Mineral Leasing Act Amendments of 1975, Interior promulgated a revised series of regulations designed to insure that leases remain in continuous operation. Holders of leases issued before August 4, 1976, were ordered to produce 2.5% of the reserves contained within their logical mining units by June 1, 1986. Leases issued after August 4, 1976 must produce 1% of the estimated reserves of their mining units within 10 years of the lease issuance. One glaring limitation of these regulations is Interior's willingness to accept unverified assessments of coal reserves submitted by leaseholders as opposed to independent agencies commissioned by the Department.

Interior's regulations also sought to discourage the speculative holding of federal coal lands by requiring advanced royalties on inactive leases. Non-producing leases in their sixth year after issuance must pay royalties based on a production schedule which would exhaust the lease's reserves in 40 years, regardless of whether the coal is being mined at that rate or not. As a further incentive to produce, the regulations allow royalty payments made by leaseholders who mine more coal than the minimum required by the production schedule to be credited against future production shortfalls.

While establishing these new regulations, the Interior Department has also begun to pave the way for regulated regional development. In January 1976 then Secretary Kleppe announced the initiation of 10 separate Environmental Impact Statements. Scheduled for final completion in 1979, these studies will cost approximately $10 million. The first of the 10, dealing with development projects in southwestern Wyoming, was issued in April 1978.

President Carter's landmark "moral equivalent of war" energy message of April 1977 ordered the Department of Interior to reshape its coal leasing bureaucracy—a task it is now undertaking. The Federal Coal Management Review Policy Committee has been established within Interior to execute Carter's mandate that leasing occur "only in areas where mining is environmentally acceptable and compatible with other land uses." Under the leadership of Guy Martin, Assistant Secretary for Land and Water Resources, the Committee is reviewing all aspects of federal leasing.

CONCLUSION

The Department of Interior has taken steps in the last several years to confront the blatant maladministration of earlier leasing efforts. New legislation, court directives, departmental regulations and continued scrutiny by a host of independent groups promise that the abuses of the past will not be repeated in new leases. CEP finds another encouraging sign in the sincere efforts of talented members of the Carter staff, who have gone beyond the token and feeble "reforms" of the Nixon Administration to begin dealing with some very difficult issues.

Still, CEP believes that the leasing program remains in deep trouble. Although many encouraging improvements have been made, they represent individual, unrelated accomplishments. The sum total of these accomplishments constitutes a new, and undeniably improved, leasing program, but it is being developed in an ad hoc fashion without any clear direction from decision makers. President Carter has refused to take any clear position on the future of federal coal leasing. Thus, the leasing policy now is a policy-by-default rather than a policy with concrete goals and objectives.

The Department of Interior has yet to grasp the most fundamental question at hand: Does it need to lease additional federal coal reserves? With 16 billion tons of coal already under lease, the arguments against new leasing are compelling. Yet, nearly all of the Department's new leasing regulations will affect only future leases. Attempts to rectify past errors, through lease ex-

46

changes, condemnations, etc., have been meager and half-hearted. The Department has not answered critics, including CEP, who believe that the past leasing program should not be allowed to mar the future of western coal development. Instead the Department has proposed programs to fit a variety of leasing policies, but left the overall policy undetermined. This amounts to a leadership void which, until filled, threatens to keep the entire leasing program hamstrung and ineffectual.

Meanwhile, even as the Department of Interior prepares study after study, makes proposal after proposal, each one conflicting with or superseding the one before, and grapples with deciding what the coal leasing policy in Washington should be, federal coal lessees in the West continue to treat public land as a negotiable and exploitable commodity.

2. STATE LEASING

INTRODUCTION

The six coal-producing states in the West were carved from vast territories acquired by the federal government in the nineteenth century. The federal government awarded each state, upon its admission to the Union, specific sections of land in every surveyed township within its borders to be owned and managed by the state government. This largesse was intended to provide the states with a resource asset, the exploitation of which would yield revenue needed to operate critical state institutions.

More than 32 million acres, constituting 6% of the total land area of the six states surveyed, are controlled by the various state governments (see Table 2-1). Nearly all of this acreage occurs in scattered, non-contiguous 640-acre sections distributed evenly throughout each state. Congress authorized this "checkerboard" pattern when, in its initial surveys of western land, it divided the territories into townships, each 36 square miles. These townships were further divided into 36 one square mile sections. Congress awarded each state sections 16 and 36. To Utah and New Mexico, Congress was particularly generous, awarding sections 2 and 32 as well. Ownership of the remaining western lands was either conferred on railroads, sold to private individuals, incorporated into Indian reservations, or retained in the federal domain.

While this checkerboarding of single-section, state-owned tracts predominates in the West, several large contiguous tracts have been amassed under state ownership. These exceptions—such as in the Bookcliffs region of Utah—resulted when a state exchanged sections of scattered land for a contiguous block elsewhere. Congress permitted these land exchanges whenever the original sections granted proved inaccessible (at the bottom of a lake, for example) or were included in a national preserve or Indian reservation.

At the time of each state's entry into the Union, Congress designated a set of public institutions, such as schools and hospitals, as beneficiaries of revenues generated from activity on state-owned lands. Congress further

TABLE 2-1

TOTAL STATE-OWNED LAND

Acres

Colorado	2,977,246
Montana	5,130,624
New Mexico	13,000,000
North Dakota	3,250,000
Utah	3,600,000
Wyoming	4,200,000
TOTAL	32,157,870

allocated to each state institution an exact number of acres, leaving the state to select the specific acreage tracts. Congress mandated that revenue derived from activity on particular acres of state-owned land must be turned over to the institution to which those acres had been assigned. Congress then directed the state to manage its lands so as to maximize the revenues accruing to these vital institutions. Upon admission into the Union, each state incorporated these Congressional directives into its state constitution.

In accordance with these provisions, each of the six western states—Colorado, Montana, New Mexico, North Dakota, Utah and Wyoming—operates a coal leasing program on its state-owned land. CEP has found that these state coal leasing programs have failed to promote orderly resource development, failed to generate substantial revenues, and failed to insure adequate protection of the environment—the hallmarks of sound land management responsibility. Many of the programmatic inadequacies reflect the constraints imposed upon the states by the Congressionally-authorized "beneficiary institution" structure, which encouraged the states to serve as trustees for a set of institutions, without considering the broader principles of resource management. As a result, in most western states the agencies responsible for the management of state-owned coal land operate in a virtual vacuum, isolated from other state agencies responsible for the implementation and enforcement of coal development policies. More often than not, the western state leasing programs have been betrayed by gross mismanagement, apathy, incompetence and servitude to corporate interests. The degree of this mismanagement, however, varies greatly from state to state.

As the federal leasing moratorium made the acquisition of federal coal leases more difficult, a mad rush to obtain state coal leases left Utah and Wyoming with virtually all of their state coal land under lease. State coal leases were issued in these and other western states for nominal filing fees or minimal cash bonuses. State coal land leases in the West are generally consolidated in the hands of a few large energy conglomerates—often with extensive oil and natural gas holdings as well—or land speculators with no mining expertise or intentions. Instead, speculators barter for their lease holdings on the assignment market, picking them up from the government, then selling them as the coal market changes. Production on state coal land, as a result, is astonishingly low: less than 1% of all valid leases have ever produced a single ton of coal.

Several western states—Montana and North Dakota, for example—have recognized the inadequacies of their leasing programs and instituted moratoria on coal leasing in the 1970s. These states have since enacted new leasing legislation and regulations which promise to rectify many of the abuses of the past. Even the revised programs in these states, however, will continue to follow guidelines which, however beneficial when Congress first mandated them, have failed to anticipate or fulfill the need for responsible resource management.

HISTORY

There are no reliable surveys of the total quantity of coal reserves underlying state-owned land in the West. The most conservative estimates range in the billions of tons. A few states have made projections based on surveys of contiguous federal or privately-owned acreage. In Wyoming, for example, roughly 1.3 million of the state's 4.2 million acres is thought to contain coal. About 900,000 of North Dakota's 3.2 million acres of state-owned land bears surface lignite. Deep bituminous seams underlie almost one-sixth of Utah's 3.6 million acres. New Mexico's coal-bearing acreage is thought to be concentrated in the northwestern part of the state bordering on the Navajo Indian reservation. Montana's most accessible coal reserves are in the southeastern portion of the state—part of the Fort Union formation. Colorado's coal reserves are scattered throughout the state, but are especially concentrated in the northwest corner.

The checkerboard pattern of state-owned coal land across the western states insures that most large coal regions are likely to include at least one section of state-owned land. The acquisition of state land by private developers is therefore vital to the organization of a logical mining unit. The efficient and economical recovery of western coal, then, is contingent upon the ability of developers to mine through state land rather than divert their operations around these sections. So while the amount of acreage controlled by these six western states is only a fraction of the acreage controlled by the federal government in the West, the development of state-owned land bears a critical importance to the coal boom forecast for the West.

To facilitate coal development on state-owned acreage, each western state has created within its government a Board of Land Commissioners to issue coal leases and regulate a coal leasing program. These Boards usually consist of a Commissioner and four other government officials, who themselves have full-time positions in the higher echelons of state government. All are usually appointed by the governor. Apart from his Land Board responsibilities, the Land Commissioner guides the state agency which actually implements and administers the leasing regulations. Table 2-2 lists each state's administrative agency designated to supervise the day-to-day operation of the leasing program.

The oldest currently valid state coal lease was issued by Utah's Division of State Lands in 1942. Coal mining on state land in the West, however, has occurred throughout the twentieth century. In fact, coal mining on Colorado and Utah state lands peaked in the 1920s and has yet to equal the level of production established then. But there are few official records of these early coal mining projects. In many western states, the pioneering miners merely filed a statement with the local government announcing an intention to extract coal. No leases were issued and no revenues were collected. The official policy in North Dakota was to allow anyone to mine coal from state land for personal use free of charge. By the end of the Depression, several western states began formalizing lease contracts with coal developers. The earliest state leases have long since been relinquished and their records destroyed.

Except for the grand entry of Exxon subsidiary, Carter Oil Co., and Mapco Inc. into the Wyoming state leasing picture in 1965 and 1966, when each of these companies acquired more than 100 leases, coal developers and land speculators paid little attention to the various state leasing programs in the 1950s and 1960s. Instead, they were busy gobbling up federal coal leases and negotiating for mammoth tracts of Indian land. But when the Department of Interior declared a moratorium on the issuance of federal coal leases in 1971, and the western Indian tribes subsequently sought invalidation of the coal leases on their land, prospective lessees turned with enthusiasm to state-owned acreage. The states have welcomed the attention—fully 50% of all

TABLE 2-2

STATE LEASE HISTORY

	Date of Earliest Lease	Current Leasing Agency
Colorado	1954	Department of Natural Resources
Montana	1963	Department of State Land
New Mexico	1960	State Land Office
North Dakota	1959	Board of University and School Land
Utah	1942	Division of State Land
Wyoming	1957	Department of Public Lands and Farm Loans

STATE LEASES ISSUED SINCE BEGINNING OF 1973

	Number	Percentage
Colorado	86	59
Montana	0	0
New Mexico	194	89
North Dakota	4	40
Utah	380	74
Wyoming	609	39
TOTAL	1,273	50

currently valid state coal leases in the West have been issued since 1973 (see Table 2-3).

Only in Montana and North Dakota have measures been taken to avert the leasing debacle that characterized the federal and Indian programs. Both states declared leasing moratoria in 1971. Montana has issued only one lease since then—a special preference right lease to Decker Coal Co. in 1971 allowing it to mine through a state tract rather than divert a costly operation around it. North Dakota's Land Board resumed leasing in 1977 with the issuance of four new coal leases under a completely refurbished program. Colorado and New Mexico, too, have promulgated new regulations designed to substantially revamp their leasing programs, but leasing in these states continued generally unabated through the 1970s. Wyoming and Utah still operate antiquated, blatantly inadequate leasing programs.

EXTENT

There are now 2,553 currently valid leases issued by the Boards of Land Commissioners in the six states examined by CEP. Of this total, 61% or 1,568 are in the state of Wyoming. Utah's 514 state coal leases rank second in the West. North Dakota's 10 state leases are the fewest (see Table 2-4). Total state-owned acreage under coal lease in the West is 2,193,630—three times the acreage under federal lease in these states. But because the checkerboarding of state land restricts most state-owned tracts to 640 acres, the average state lease—859 acres is much smaller than western federal or Indian leases, which average 1,502 and a huge 24,000 acres respectively.

There are some exceptions to the rule of small state leases, however. CEP found nine leases in Utah whose acreage exceeded the statutory limit of 2,540 acres. Four mammoth leases, each larger than 10,000 acres, have been issued in Colorado. The existence of these four leases explains why more than twice as much state acreage is under lease in Colorado as New Mexico, though 32% fewer leases have been issued there.

TABLE 2-4

EXTENT OF STATE LEASING

	Number of Leases	Number of Acres	Average Lease Size (Acres)
Colorado	147	252,199	1,715
Montana	96	51,947	514
New Mexico	218	106,860	490
North Dakota	10	3,838	384
Utah	514	543,557	1,057
Wyoming	1,568	1,235,229	788
TOTAL	2,553	2,193,630	859

ISSUANCE PROCEDURES

All state coal leases have been issued according to one of five methods: open land, preference right, competition, negotiation or segregation. Open land leasing allows any corporation or individual qualified to conduct business within the state to receive a coal lease simply by demonstrating that the state-owned tract has not already been leased for coal development. Under preference right procedures, the prospective lessee first applies for a prospecting permit which may be converted into a lease if the prospecting uncovers coal. Competitive leasing awards the lease to the applicant who offers the most generous one-time cash bonus payment. Under the provisions of negotiated leasing, the state leasing agency and prospective lessees agree on precise bonus payments, rental and royalty rates. Segregated leases are issued by splitting an existing lease into two or more components, the sum of whose acreage equals that of the original lease. Table 2-5 illustrates the distribution of all valid coal leases within the six states according to lease issuance method.

Open Land

The open land lease issuance procedure requires only minimal participation by the state Land Board. States that follow this procedure direct the leasing agency to prepare land maps which denote each section owned by the state and maintain records of all tracts leased to the private sector. A coal lease applicant may review these maps and records at any time. An applicant who finds state acreage not yet leased for coal development may submit a simple application form requesting the lease. If, in fact, the tract is owned by the state and not covered by a valid lease, the applicant *must* be issued the lease upon payment of a small filing fee—usually $15. At no time is any lease applicant required to prove the existence or even suspected existence of coal under the land. Neither must any proof of intention or capacity to mine coal be

TABLE 2-5

STATE LEASE ISSUANCE PROCEDURES

	Open Land	Competition	Preference Right	Negotiation	Segregation
Colorado	92	46	0	0	9
Montana	31	64	1	0	0
New Mexico	0	194	24	0	0
North Dakota	0	6	0	4	0
Utah	471	23	0	0	20
Wyoming	1,568	0	0	0	0
TOTAL	2,162	333	25	4	29

registered. The state only requires that the applicant be in compliance with the laws of the state and qualified to conduct business within it.

There could not be a system more conducive to the random giveaway of public resources. Yet the open land system is clearly the historically preferred procedure among state governments, having been responsible for the issuance of 85% of all outstanding coal leases and 91% of all leased state acreage. All 1,568 state leases in Wyoming, and 471 of the 514 leases issued by the Utah Division of State Lands, have been issued according to the provisions of open land leasing. Colorado and Montana both utilized this procedure in the early days of their leasing programs, but abandoned its use in the 1970s.

Wyoming has employed a lottery system for the reissuance of open land leases available again for leasing due to a cancellation, termination or relinquishment of the original lease. Technically, this lottery constitutes a separate issuance procedure. But because Wyoming maintains no records of the number of leases issued by lottery, and because the non-competitive nature of the lottery so closely mirrors that of the open land system it is intended to complement, CEP has included these lottery leases as open land leases in Wyoming.

When a state tract becomes available for re-leasing, Wyoming's Department of Public Lands and Farm Loans advertises its availability. Prospective lessees may then file an application to obtain the land. Within 10 to 30 days after first notification of the lease sale, the lottery is held. Instead of submitting a sealed bid as is the practice for second leasing in Utah, applicants place a card in a metal drum. The drum is revolved and a card is withdrawn. The applicant whose name appears on the card wins the lease upon payment of a $15 filing fee.

Preference Right

Preference right leasing is also a non-competitive system which authorizes the issuance of coal leases upon payment of only a minimal filing fee. As with the open land system, a lease applicant first identifies a tract of unleased state-owned land. The next step is to apply for a prospecting permit, which confers the right to explore for coal for a one-to-two year period. An applicant who can then supply the state leasing office with geological data confirming or suggesting the presence of coal in commercially mineable quantities there may request a lease. If the leasing agency verifies or accepts the validity of the applicant's data, it must issue the lease non-competitively upon payment of a minimal filing fee—again usually $15.

New Mexico's Land Board has been the only state leasing agency in the West to employ preference right leasing with any regularity. Of the 218 valid state coal leases there, 24 have been issued via preference right leasing. The last preference right lease awarded in New Mexico was in 1973. The New Mexico Land Office told CEP that no more preference right leases will be issued there, however. Montana has been the only other state to utilize preference right procedures. In 1973 one preference right lease was issued there to Decker Coal Co. to enable mining operations on adjacent federal and private land to expand onto a small state tract.

Competition

Competitive leasing procedures, of which several varieties have been employed in the western states, have accounted for the issuance of 333 leases spanning 185,748 acres. This second most popular leasing procedure in the West has been utilized by the Land Board in every state surveyed except Wyoming. Only in New Mexico, where 194 of the 218 valid leases have been issued competitively, however, has competitive leasing accounted for the majority of leases issued to date.

The different competitive leasing systems found in the western states do share certain similarities. All lease tracts offered for competitive leasing are initially "nominated" by the corporations or individuals interested in obtaining them. The state leasing agency then notifies other major mining interests of the land tract's availability for competitive leasing. All competitive state leasing programs have relied upon cash bonus bidding at public meetings to determine the winning bidder. Cash bonus bidding awards the lease to the bidder offering the highest one-time cash bonus, payable in full prior to the issuance of the lease.

Within these parameters, several competitive leasing variations have been utilized. Colorado and Montana at one point permitted oral bidding at public lease auctions—a procedure now abandoned in both states. Sealed bidding, in which prospective lessees submit their cash bonus bids in writing prior to the lease sale, has been a more widely-used procedure. On the day of the lease sale, the state's representative opens the sealed bids one at a time, reads them aloud, and offers the lease to the highest bidder. Now utilized in Colorado, Montana, New Mexico and Utah, sealed competitive bidding has accounted for 84% of all leases issued competitively.

Utah's Division of State Land utilizes the sealed bidding procedure only when a state tract already leased non-competitively becomes available again for leasing because of lease termination, cancellation or relinquishment. When this occurs, the state notifies its mailing list of coal interests that the land is again available for leasing. Applicants are accorded a 15-day simultaneous filing period to submit sealed bids. A competitive sale is then held and the bids are opened. Twenty-three state leases covering roughly 3% of all land under lease have been issued a second time under this competitive structure.

In the archives of the various leasing agencies in the West, there is complete data on 182 of the 333 leases issued competitively. CEP found that the competitive leasing programs in the western states—those that authorize oral bidding as well as those that authorize sealed bidding—have been anything but competitive. All of the procedures have repeatedly failed to generate competition for the lease acreage or raise revenues for the state. CEP found that 13 of the 46 leases issued at oral auction in Colorado stipulated no cash bonus, indicating that only one bidder appeared. Sealed bidding in New Mexico has proven no more effective. When the State Land Office there offered 299 tracts for competitive leasing in November 1977, sealed bids were received for only 90. More than one bidder submitted a sealed offer for only 19, or 21% of the leases issued "competitively" that day. So, too, for sealed bidding in North Dakota. The six leases issued competitively through 1971 there attracted an average bid of only $1.16 per acre. More than one bidder appeared at only two of these sales.

Crippled by lack of any significant competition, the various competitive leasing programs have produced a total cash bonus payment of $1,125,645—an average of only $6.06 for every acre of state-owned land leased competitively in the West. Only one bidder offered a cash bonus payment for 131, or 72% of these lease tracts. The winning and only bids submitted here averaged $3.72 per acre. Two bidders—the absolute minimum for any semblance of a competitive atmosphere—appeared at 40, or 22% of these lease sales. These 40 leases were issued for an average bonus of $11.21 per acre. Three bidders appeared at only 11 lease sales constituting 6% of all leases for which complete data was available. The average winning bid was $10.26 per acre for these tracts. At no lease sale did more than three bidders appear to offer a cash bonus.

At the other extreme, North Dakota graciously fulfilled one applicant's request for a free state coal lease.

Segregation

Leasing provisions in Colorado and Utah permit a leaseholder to petition the state leasing agency to divide an already valid lease into two or more distinct entities. The original lessee may retain ownership of any, all or none of the resulting leases. Usually, ownership of at least one of the new leases is transferred to another party.

This process of segregating out a lease is not a true lease issuance procedure because it creates no new lease acreage. Still, it does produce a new lease contract upon payment of a small segregation fee to cover the cost of paper work involved in its preparation.

Negotiation

Negotiation of lease terms requires greater participation from the state leasing agency than any other lease issuance procedure. The process begins when a prospective lessee proposes a lease package—including suggested cash bonus payment, rental rates and royalties—to the state. State officials then join representatives of the applicant in finalizing terms. If an agreement is reached, the state publishes the proposed lease deal.A competitive auction is then held at which any other party may best the proposed terms and obtain the lease for itself.

To date, only North Dakota has negotiated lease terms with prospective lessees. Terms of four leases issued in December 1977 have been negotiated. Although North Dakota officials had hoped that lease negotiation would raise more revenues for the state than did the old competitive leasing program, each of the four negotiated leases was accompanied by a cash bonus payment of only $1 per acre. The lessees will pay royalty rates of 10% of the gross value of the coal, however—a fairly high royalty rate, although not the highest CEP encountered. Rental rates too will escalate from $1 to $5 per acre per year.

ASSIGNMENT

The barriers to entry in the coal leasing market are extremely low. Ninety-one per cent of all state land now under lease has been leased according to open land provisions for a nominal filing fee—usually $15. North Dakota is the only western state which requires lease applicants to demonstrate the capacity and the intent to mine coal. Elsewhere in the West it is open season for speculators and others with neither the capital, expertise nor interest in mining coal to acquire coal leases.

Each western state examined invites these speculators into its coal leasing program by permitting the selling and bartering of leases on the open market. State officials throughout the West conceded to CEP that these lease "assignments" are accompanied by sizeable monetary transactions. Except for Colorado, none of the western state governments share in the cash transaction, nor are they even permitted to ask the amount of cash changing hands. Overriding royalties requiring the assignee to pay a fee to the assignor based on the quantity of coal extracted must be reported to the state, however. State leasing agencies may deny an assignment request only if they determine the overriding royalty to be so high as to inhibit eventual coal production. An assignment request has never been denied in the western states.

STATE LEASE ASSIGNMENTS

	Number of Leases Assigned	% of Leases
Colorado	48	33
Montana	40	42
New Mexico	43	20
North Dakota	3	30
Utah	315	61
Wyoming	423	27
TOTAL	872	34

The extent to which lessees avail themselves of the assignment option is startling—and still on the rise. Almost half of all valid state coal leases in the West issued prior to 1973 have been assigned to their present owner since then. At one time or another, 34% of all western state coal leases have been assigned (see Table 2-6).Assignments have occurred in every western state. An amazing 61% of all Utah state leases have changed hands at least once. New Mexico's assignment rate of 20% is the lowest in the West, but the vast majority of New Mexico's leases have been issued in the last two years, leaving little time for assignment transactions to occur.

CEP believes that assignment provisions do more than any other single leasing policy to promote lease speculation and hamper orderly land and resource management. As state leasing agencies continue virtually to give away their coal land to speculators who in turn sell their lease rights for sizeable profits, the assignment market has become the true competitive arena for the selling of public coal. Unfortunately, the public does not benefit from it.

OWNERSHIP

The ownership of state coal leases in the West is scattered among 400 individuals and corporations. Considerable concentration of lease ownership exists, however. The top five leaseholders in each of the western states control an average of 45% of the land under lease. In no state do the top five leaseholders control less than 43% of the acreage under lease in the state (see Table 2-7).

Mapco Inc., a fast growing energy conglomerate with coal, oil and other energy resource holdings, is the largest state coal leaseholder in the West (see Table 2-8). In the six states surveyed, Mapco controls 154 leases covering

TABLE 2-7

OWNERSHIP OF STATE LEASES

	Number of Leaseholders	Acres Controlled by Top 5 Leaseholders	% of Acres Controlled By Top 5
Colorado	63	111,096	44
Montana	16	34,018	65
New Mexico	33	51,146	48
North Dakota	5	3,838	100
Utah	123	259,407	48
Wyoming	298	529,062	43
TOTAL		988,567	45

TABLE 2-8

TOP FIVE STATE LEASEHOLDERS

Leaseholder	Number of Leases	Acreage Under Lease
Mapco	154	168,257
Carter Oil Co.	175	165,786
Ark Land Co.	98	98,124
Coastal States Energy	61	92,277
R.A. Haynesworth	258	82,059
TOTAL	746	606,503

168,257 acres. Its most extensive holdings are in Wyoming, where it is the second largest leaseholder, and Colorado where it is the largest.

Exxon subsidiary Carter Oil Co. is the second largest holder of state coal leases in the West. It is the top leaseholder in Wyoming and second largest in Montana. Its 175 leases span 165,786 acres.

Ark Land Co., a wholly owned subsidiary of Arch Minerals Corp.—itself jointly owned by Ashland Oil and Hunt Enterprises—is the third largest state leaseholder. Its holdings also are most extensive in Wyoming; however, Ark Land controls lease acreage in every western state except North Dakota and Utah.

Coastal States Energy Co., a Houston-based unit of Coastal States Gas Co.—also with large oil and gas holdings—is the West's fourth largest state coal leaseholder. The top leaseholder in Utah and the fourth largest in New Mexico, Coastal States controls 92,277 acres throughout the West.

R.A. Haynesworth, a land speculator who has partial or complete control of 258 leases in Wyoming, is the fifth largest holder of state land in the West. Haynesworth has acquired all his leases since 1971. He has outright ownership of only seven leases. Of the remaining 251 which bear his name, 239 are held in joint ownership with one other party, 10 are in tri-party ownership, and two are part of a four-person lease ownership team. Each of Haynesworth's leases was acquired for the $15 filing fee required by open land leasing procedures, rather than by purchasing leases on the assignment market. This is hardly surprising, however, since he himself is a lease broker with an eye towards profiting from the bartering of land—in this case, coal land. Haynesworth's base is Cheyenne, Wyoming, but CEP found that the addresses of his co-lessees are scattered across the country, from Marietta, Ohio to Hawaii. CEP questions whether Haynesworth or his elusive co-lessees will ever initiate a mining venture on Wyoming land.

Apart from Exxon's subsidiary, Carter Oil, other major oil companies are heavily involved in state leasing programs in the West. Continental Oil's subsidiary, Consolidation Coal Co., is the number one leaseholder in Montana and the number three leaseholder in Colorado. Both Gulf and Mobil Oil Cos. also appear on the list of the top five leaseholders in at least one western state.

Investor-owned utilities are prominent among the list of major leaseholders of western state coal land. American Electric Power Co.'s Franklin Real Estate subsidiary is the second largest Colorado leaseholder, and also has extensive holdings in Utah. Decker Coal Co.—a joint venture of Peter Kiewit Sons and Pacific Power & Light—is the fifth largest holder in Montana. Knife River Coal Co., a Montana/Dakota Utilities subsidiary, owns extensive state coal lands in North Dakota. The Salt River Project, a consortium that includes several southwestern utilities, is the second largest state leaseholder in New Mexico.

These leading holders of state coal land are often involved in the federal and Indian leasing programs as well. Consolidation Coal, for example, is the second largest federal coal leaseholder. It also has one-half interest in the CONPASO venture on a huge lease on the Navajo Reservation in New Mexico. Utah International, the second largest state leaseholder in Utah, holds over 55,000 acres in federal and Indian leases. Peabody Coal, the third largest state leaseholder in Montana, holds three of ten Indian leases and is the largest holder of federal coal leases in the country.

PRODUCTION

Only 19 of the 2,553 valid coal leases produced any coal in 1977, thus accounting for only 0.7% of all outstanding leases (see Table 2-9). Moreover, only 23 state coal leases in the West have ever produced a single ton of coal.

No New Mexico state lessee has ever produced any coal on any of the 218 valid coal leases there. In Wyoming, a mere 0.001% of all state leases are now

NUMBER OF COAL PRODUCING STATE LEASES IN 1977

	Number of Producing Leases	% Producing
Colorado	3	2
Montana	1	1
New Mexico	0	0
North Dakota	5	50
Utah	7	2
Wyoming	3	0.001
TOTAL	19	0.7

TABLE 2-10

1977 COAL PRODUCTION AND ROYALTIES FROM STATE LAND

	Tons Produced	Royalty	Royalty/Ton
Colorado	218,006	$119,279	54.7¢
Montana	5,115,418	895,198	17.5
New Mexico	0	0	0
North Dakota	1,267,893	132,563	10.4
Utah	320,028	44,909	14.0
Wyoming	730,000	124,000	17.0
TOTAL	6,651,345	$1,315,949	18.1¢

active. Only in North Dakota are more than 2% of existing state leases currently producing coal. Because of strict leasing regulations there, which only permit the issuance of leases to applicants who demonstrate the capacity and intent to mine coal, five of the 10 valid leases are now in production. Firm plans have been filed by the other five leaseholders to bring their land into production.

Total output from the 19 active leases in 1977 was 6,651,345 tons—roughly 12% of the 1977 output on western federal land (see Table 2-10). Decker Coal Co. contributed more than five million tons to this total from its Decker mine situated partially on state land in southeastern Montana. North Dakota's five active leases generated 1.2 million tons of coal last year, making that Northern Plains state the second largest coal producing state in the West, even though it has the fewest outstanding leases and the fewest acres under lease in the region.

Although production on state lands remains extremely low, state leased reserves contribute to the output from some of the largest mines in the country. The Decker mine which so dominated state coal production in 1977 was the second largest mine in the country that year. The Sunnyside and Horse Canyon mines in Utah which supply metalurgical coal to U.S. Steel and Kaiser Steel are also partially situated on state land. Consolidation Coal Co.'s active lease tract in North Dakota is part of its Glenharold mine, the largest coal mine in the state. Production on state land is only a small part of the total output of these and other mines. Most of the production comes from adjoining federally-leased or privately-owned reserves.

REVENUE

The Constitution in each western state directs the state leasing agency to manage its state-owned land so as to foster commercial activity and maximize revenues accruing to critical state institutions. State coal leasing generates four sources of revenue: from the issuance of coal leases, from rental payments, from royalties based on production, and from advance royalties paid in lieu of production. Royalty payments eliminate the lessee's obligation to pay advanced royalties. In most states, rental payments may be credited in full against actual or advanced royalties. This layering of credits has contributed to the failure of the western state leasing programs to generate a flow of revenue adequate to compensate the state for its land losses.

Lease Issuance Revenue

The six state leasing agencies have collected $1,197,194 from the issuance of all currently valid state coal leases—an average of $469 per lease or 65¢ per acre (see Table 2-11).

Bonus bid payments accompanying the issuance of the 333 competitive leases have constituted the largest source of lease issuance income to the states. Every western state except Wyoming has at one time or another employed competitive bidding. New Mexico, which has operated the most aggressive competitive leasing program to date, has garnered nearly 75% of all revenues accruing to the western states from competitive lease issuance. Even so, New Mexico has received just $7.41 per acre of land leased competitively. North Dakota has received the least amount of cash bonus payments, averaging just $1.16 per acre.

Open land leasing has brought $47,653 to state leasing agencies—an average of 2¢ per acre. Most states charge a mere $10 or $15 for the right to obtain an open land lease. Colorado adds 10¢ per acre to its basic filing fee of $10. The filing fee in New Mexico graduates from $10 to $25 according to the size of the lease.

TABLE 2-11

REVENUES FROM STATE LEASE SALES
(dollars)

	Open Land	Preference Right	Compe-tition	Segre-gation	Negoti-ation	Ex-changes	Total
Colorado	21,468	0	102,549	90	0	0	124,107
Montana	310	1,064	207,204	0	0	77,118	285,696
New Mexico	0	920	724,572	0	0	0	725,492
North Dakota	0	0	3,637	0	720	0	4,357
Utah	2,355	0	31,527	140	0	0	34,022
Wyoming	23,520	0	0	0	0	0	23,520
TOTAL	47,653	1,984	1,069,489	230	720	77,118	1,197,194

Preference right leasing programs in Montana and New Mexico have generated $1,984—an average of 22¢ per acre. Decker Coal Co. contributed most of this revenue, $1,064, for the right to obtain a small preference right lease during Montana's state leasing moratorium. Preference right leasing in New Mexico has yielded just $38.33 per lease.

Negotiated leasing in North Dakota has produced $720 for the issuance of four leases. The State Board of University and School Lands justifies the low cash bonus payments it accepted, $1 per acre, by pointing to high rental and royalty rates included in the negotiated lease terms. CEP has found the graduated rental rates and royalty rates of 10% of the gross value of the coal to be among the higher rates effective in the West, but not high enough to off-set these nominal bonus payments.

Segregation of existing leases has brought in $140 and $90 to the Utah and Colorado leasing agencies respectively. Colorado assesses a $10 segregation fee while Utah's is $7. In no other state have leases been segregated.

Lease exchanges in Montana have constituted the final source of lease issuance revenue. In 1967, after 19 leases had already been issued, Montana revamped its leasing program, promulgated new standard lease terms, and required holders of existing leases to exchange their contracts for new leases incorporating the new lease terms. Each leaseholder was charged an exchange fee, varying with the size of the lease. Revenue collected from these 19 ex-changes totalled $77,118.

Rent

Every western state requires leaseholders to pay rent at least until production begins on the leased tract. These payments represent the cost of maintaining a non-producing state coal lease. Current rental requirements place only a minimal financial burden on leaseholders, thereby encouraging speculation rather than production on state acreage.

Several different rental structures are employed in the West, varying from state to state and with the age of the lease. Most valid leases authorize rental rates between 25¢ and $1 per acre per year. Several Montana leases contain rental rates that begin at $1 an acre but jump to $5 per acre per year beginning with the sixth year of the lease. The four negotiated leases in North Dakota require lessees to pay $1 per acre per year for the first five years, $3 per acre per year for years six through eight, and $5 per acre thereafter.

In Fiscal 1977, the six state leasing agencies in the West collected $904,190 in rent payments—an average of only 45¢ per acre for all state land under lease (see Table 2-12).

Royalties

Once coal development begins on state land, the lessee is required to pay a fee based on tonnage extracted. In 1977 lessees from the five states supporting mining operations on state coal land—no coal mining has ever occurred on New Mexico state land—paid $1,315,949 in royalties. This payment represents an average rate of 18¢ per ton, or, with western coal selling for roughly $15 per ton, 1.3% of the value of the coal (see Tables 2-10 and 2-12).

Western states employ either a fixed or variable royalty schedule, although the rates themselves vary considerably from state to state. Fixed royalties require lessees to pay a fixed fee—usually between 5¢ and 20¢—on each ton of coal produced regardless of the value of the coal. Every state examined once charged lessees a fixed royalty. Consequently, most outstanding leases authorize this kind of payment. Utah is the only western state examined which still issues leases authorizing fixed royalties. Every other state has switched to a system which requires lessees to pay the state a percentage of the value of the coal mined. Wyoming is now charging new lessees 8%; Montana and North Dakota are charging 10%; Colorado charges underground mine operators 8% and strip mine operators 12.5%; New Mexico charges a flat 12.5%. In no state does the variable royalty rate surpass the 12.5% minimum established by the Mineral Leasing Act Amendments of 1975 for the federal leasing program.

Advanced Royalties

Advanced royalties are payments lessees make to the state leasing agency if, after a specified period of time, their leases remain inactive. The purpose of advanced royalties is more to encourage leaseholders to develop state resources than to generate additional revenue to the state.

CEP has found that the advanced royalty rates vary in the West, but they are never high enough to provide sufficient incentive for leaseholders to open mining operations on their leased land. Montana lessees are required to pay a flat $50 per lease per year in advanced royalties. No advanced royalties are charged in Utah or Wyoming until the initial 10-year lease terms expire;

1977 REVENUES FROM STATE LEASEHOLDERS
(dollars)

	Rent	Advanced Royalty	Actual Royalty	Total
Colorado	145,569	55,160	119,279	320,008
Montana	87,488	4,750	895,198	987,436
New Mexico	39,511	328,344	0	367,855
North Dakota	3,838	0	132,563	136,401
Utah	291,784	46,300	44,909	382,993
Wyoming	336,000	900,000	124,000	1,360,000
TOTAL	904,190	1,334,554	1,315,949	3,554,693

thereafter, lessees must pay $1 per acre per year to renew their coal leases. New Mexico charges $3 per acre in the first year of the lease, $4 per acre in the second year, and $5 per acre thereafter. In Colorado, lessees are required to pay $6 per acre per year beginning in the sixth year of the lease; this amount rises each year by $1 per acre until the tenth year. This $10 per acre per year rate is the highest in the West. North Dakota, alone among the western states, charges no advanced royalties.

In Fiscal 1977, the five state leasing agencies which assess advanced royalties of holders of inactive leases collected $1,334,554—an average advanced royalty rate of 61¢ per acre (see Table 2-12). The fact that total advanced royalty collections surpassed actual royalties last year further underscores the pervasive inactivity on state-owned coal land in the West.

Total Revenues

CEP's survey of revenues derived from rent, royalty and advanced royalty payments accruing to the western state leasing agencies reveals that 1977 collections totalled $3,554,693. Advanced royalties accounted for slightly more income to the states than actual royalties. Rent payments contributed roughly one-fourth of the total.

Wyoming collected more revenue than any other western state last year—a reflection of the extensive acreage under lease, not of the stiffness of the rates there. Most of Wyoming's income derived from advanced royalty payments. Montana, with only 3% of the total state land under lease in the West, ranks second in 1977 cash receipts from coal leasing. Most of Montana's income came from royalties paid by Decker Coal Co. for production on one lease tract. Nearly all of Utah's rent, royalty and advanced royalty income came in the form of rent payments, with rates ranging as low as 25¢ per acre.

The absence of coal production on New Mexico's leased state lands insured that the distribution of lease income there would be lopsided in favor of advanced royalties.

BENEFICIARY INSTITUTIONS

When an Indian tribe in the West decides to lease some of its coal reserves, it does so out of the conviction that leasing is in the best interest of the whole tribe. Similarly, the federal government's leasing programs seek to promote orderly and timely resource development, obtain a fair return for the sale of public resources, and protect the environment—objectives that Congress has determined to be in the best interest of the people of the United States. Promotion of the general welfare of its citizens is not, however, the objective that guides the various western states in operating their leasing programs. Constitutional law requires the states to manage their land so as to promote the specific and direct welfare of a finite set of institutions within the state. State leasing programs are directed toward maximizing the revenues accruing to these institutions. Consequently, state leasing agencies serve as fiducial representatives, managing accounts and investments for their beneficiaries. They are not, in any sense, land or resource management agencies.

Medieval English common law established the tradition of alloting public land for private development to benefit the King—a tradition the U.S. Congress continued in the early days of the republic. In a series of laws promulgated in the nineteenth and early twentieth centuries, Congress granted to the newly formed state governments in the West control over a small percentage of lands within their states. As part of the Enabling Act preceding each state's entry into the Union, the federal government further designated a list of then critical institutions within the state as beneficiaries of any activities generating revenue on specific acres of state land. It alloted precise acreages to each institution, the vast majority, 85-90%, going to the common schools. The remaining land was assigned to a long list of small institutions including prisons, hospitals, universities, insane asylums and schools for crippled and blind children. CEP found 82 institutions now receiving revenues from state land in the West.

One of the earliest acts of each newly-admitted state's government was the selection of specific land tracts for each institution, and development of a system to insure that all revenues derived from activity on a section of state land benefit the specific institution assigned to it.

Each western state leasing agency manages its revenues in a similiar manner. "Earned" income, derived from activities which do not deplete the value of the land, is immediately allocated to each institution and becomes part of its operating budget. Earned income produced by the state coal leasing programs includes rent and cash bonus payments. Royalties and advanced royalties are considered revenues derived from activities which reduce the value of the land

and so are allocated differently. To maintain the value of the land as an asset to each institution, the states do not permit royalty collections to be spent. Instead, these monies are deposited into State Permanent Funds which are invested in stocks, bonds and other common investment alternatives.

Each year, the various western states allocate to the different institutions the interest income earned from the investments of the Permanent Funds. Although all revenues deposited in the funds are commingled, the interest income is divided among the institutions in proportion to the contribution of revenues from land assigned to each institution. For example, if 80% of the Permanent Fund's principle is derived from activities on common school land, then the common schools are assigned 80% of the fund's interest annually. The precise proportions of fund dispersement among the institutions are adjusted on a monthly, quarterly or yearly basis.

The principle of the Permanent Funds is held in perpetuity for the institutions. It is intended to represent the value of assets lost to the state because of commercial activity on its lands. As the pace of development on state land has increased throughout the twentieth century, so too have the assets of the various Permanent Funds. State land in New Mexico contains many valuable natural gas and oil deposits, the development of which has brought its Permanent Fund balance up to $736 million since 1910—the highest fund balance in the West. Colorado's Permanent Fund, with a balance of $59 million, is the smallest. In total, the six state Permanent Funds have amassed $1.3 billion held in trust for the beneficiary institutions (see Table 2-13).

Coal leasing activity has contributed only minimally to the state Permanent Funds, its revenues constituting a few percentage points of the fund balance in each western state. This stands to change, however, especially in states like North Dakota and Wyoming, which anticipate expanded coal development on state land with a concurrent decline in oil and gas exploration. North Dakota officials expect the value of the Permanent Fund

TABLE 2-13

STATE PERMANENT FUNDS

	Balance (Million dollars)	Number of Institutions
Colorado	59.0	9
Montana	79.0	12
New Mexico	736.0	22
North Dakota	80.6	14
Utah	121.8	12
Wyoming	224.0	13
TOTAL	1,300.4	82

there to jump from $80 million to over $100 million in the next few years primarily because of coal development.

In their zeal to maximize the revenues accruing to these Permanent Funds and beneficiary institutions, western state leasing agencies have indiscriminately promoted any and all industrial activity on state land, at the expense of sound land or resource planning. The more activity, the greater the benefits that accrue to their institutions. Following this principle, the state leasing agencies have persisted in issuing coal leases prior to the emergence of a competitive interest in the land. In their narrow field of vision, the more leases issued, the more rent and advanced royalty payments collected. And a lease issued today is money collected today: the state leasing agencies in the West have uniformly focused on the short-term at the expense of the long. Lack of foresight in Utah and Wyoming has left nearly all state-owned land there leased for giveaway terms prior to the emergence of a competitive market for the land—a betrayal of the long-term interests of the institutions there. Leasing agencies throughout the West have yet to display the patience required to withhold state land from leasing until market changes or mineral discoveries enhance its value.

Similarly, the constitutional mandate to maximize revenues to all Congressionally-designated institutions has prevented state leasing agencies from creating a comprehensive regional development policy for its resources. Sound land management tenets would direct state agencies to select a few prime sites for development and lease them competitively for sizeable cash bonus payments. Because this procedure would benefit some institutions and not others, it has not been implemented. Instead, the western state leasing agencies have leased any land to any applicant. Leasing in this manner has insured that revenues remain minimal but adequately distributed among the institutions.

Still, the randomness of this system can be neither concealed nor excused. If coal is uncovered on a section of land assigned 100 years ago to a penal institution, only that institution is lucky enough to garner any revenue from its exploitation. If the land assigned to an insane asylum proves barren, it receives no revenue no matter how worthy its budget needs.

While few question the need to adequately fund the Utah School for the Deaf or the Wyoming Hospital for Crippled Children—although some may have doubts concerning the social contributions of the New Mexico Military Institute—the whole concept of making these institutions' funding contingent on resource development is suspect. Conversely, the flow of revenue from coal-related activity to the operating budgets of state institutions precludes the option of using these funds to reduce the adverse social and environmental impact of coal development.

CEP has found that the operation of state coal leasing programs in the West for the exclusive benefit of a finite set of institutions picked by Congress 100 years ago has resulted in the uncoordinated and unsystematic transfer of millions of acres and billions of tons of public resources into the private sphere. Because this strange system was established by Congressional mandate and incorporated into each state's constitution, formidable obstacles attend its reform.

States may, however, alter their interpretation of the beneficiary institution system. Passage of the 1975 Land Consolidation Act in Wyoming suggests that states may possess greater latitude than they exercise in their leasing programs. This bill eliminated the annual allocations of rent collections to several small institutional members of the trust. The state argued that the contributions of rent for leased state land to the budgets of these institutions were too insignificant to merit the expensive bookkeeping required to maintain the separate accounts. Rent is now deposited in the General Fund, to be allocated to the institutions as part of the general state support they enjoy. To date the statute has not been challenged by the U.S. Justice Dept.

Similarly, North Dakota's new leasing program reinterprets its responsibility away from the maximization of revenues towards the maximization of "values." The state has adopted a "go slow" leasing strategy. Only companies that can guarantee mining operations on state land shortly after obtaining a lease contract will be issued leases there.

Elsewhere in the West, the beneficiary institution structure and its conventional interpretations appear firmly entrenched. State officials generally seemed not only too lethargic to test the limits of their latitude, but aghast at the prospect of change.

CONCLUSION

Strategic location and extensive reserves insure that state-owned land will have a vital impact on coal development in the West. A competently administered coal leasing program could provide each of the western states with a substantial revenue flow. Moreover, the experience gained from developing and implementing adequate state leasing regulations could spill over and affect federal and Indian leasing decisions in the West.

Western state leasing agencies have only recently begun to approach their potential for resource management and revenue generation. In the last two years, each of the western states except Utah and Wyoming has substantially revamped its leasing program. In every state, the new leasing programs will greatly increase revenues from lease sales through strengthened competitive or negotiated leasing procedures. Rental and royalty rates have also been raised throughout the West. Montana and North Dakota have incorporated vigorous "diligent production" requirements and other provisions expanding the state's role in limiting speculation and determining the scale of coal development on state land. Additionally, the environmental reclamation potential of prospective lease sites and lease packages will be reviewed in these two states prior to the issuance of new leases. North Dakota, whose new leasing program is exemplary among the western states, has made future leasing of coal land contingent on the prior acquisition of adjacent federal or private land and the development of a coal mining plan which requires the exploitation of state reserves.

The new leasing provisions in these states are not retroactive: they apply only to leases issued since their promulgation. The vast majority of the state coal leases in the West contain the old terms which have proven so inadequate. Not until these leases come up for renewal—lease terms are five years in New Mexico, 20 years in Montana, and 10 years elsewhere in the West—will the western state leasing agencies have the opportunity to upgrade their provisions. Whether the state agencies will aggressively seek to incorporate the newer, tougher regulations in renewed leases or simply rubberstamp the old contracts for an additional term, however, remains to be seen.

The new regulations will have varying effects on the future of state coal leasing in the West. In Colorado, Montana and North Dakota—where substantial state coal-bearing acreage remains unleased—the shape of state leasing programs may be profoundly altered. North Dakota, for example, has leased only a tiny fraction of its 900,000 acres of state land thought to contain lignite. Montana has leased only 52,000 acres of its even more extensive resource-bearing acreage.

In other states, the prospects for change are decidedly bleak. Wyoming has already leased all of its state land thought to contain coal. Utah has leased 94% of its state coal lands. Both of these states continue to operate inadequate leasing programs that betray rather than serve the long-term interests of the state. In New Mexico, where limited strides have been taken to increase the revenues accruing to the state from coal leasing, 85% of the state-owned land in the Star Lake-Bisti region, the state's prime coal basin, has been leased for coal development. In these regions, state agencies no longer operate lease issuance programs in any real sense, but rather, lease maintenance programs. Until existing coal leases come up for renewal, agencies in these states are hamstrung to control development or increase financial return to the state.

Coal output too will vary from state to state in the West. In Montana, Wyoming and, to a lesser extent, New Mexico, state land output should mirror the rapid expansion of production forecast for federal, Indian and private operations there. Expansion of production on state land in North Dakota depends in large part on the commercialization of coal gasification technologies. With gasification, a huge upsurge in coal mining on state as well as federal and private land will occur. Without it, slow steady growth is the forecast. In Colorado, CEP uncovered four state leases each spanning more than 10,000 acres of contiguous state land, each capable of supporting a major mining operation that could cause state output to jump substantially. Utah also controls large contiguous state tracts in the Bookcliffs region. Leasing and diligent production there could occasion similar growth of output.

Whatever the actual rate of development in the years to come, CEP found all state leasing agencies in the six western states inadequately equipped to deal with the tasks at hand—the promotion of orderly resource development, the generation of revenues reflecting fair market values, and the protection of the environment. In each state leasing agency visited, CEP found incomplete and often contradictory data haphazardly tucked away, inaccessible to the public. Most state leasing officials conceived their roles in very limited terms—to barter public resources to any private party. Few sensed the inadequacies of their programs or anticipated the need for reform. In most states CEP questioned officials of other governmental agencies more directly concerned with the regulation of coal development and its impact. Often these officials characterized the state leasing agencies as the laughingstocks of the mineral development bureaucracy. With the future of state leasing in such hands, the prospects for significant and much needed changes appear dim.

Over the past few years the states have been requesting funds from the federal government to help them cover the costs of the adverse impact of rapid coal development. The states claim they are victims of the Department of Interior's mismanagement of the federal leasing program. While CEP's evaluation of the federal program substantiates the states' claim, it is also true that the states have done little better, if not worse in some respects, in managing their own land. Together the federal and state governments have contributed to an overall poor coal leasing situation in the U.S.

3. INDIAN LEASING

INTRODUCTION

The early white pioneers stretched westward the boundaries of the new American republic, ultimately at the expense of the Native American tribes who had roamed the great plains and prairies. Challenging the white man's power to subdue the land, these tribes were eventually subdued themselves and pushed onto pockets of land usually worthless for agriculture or other traditional livelihoods.

There most tribes have remained amidst poverty and powerlessness for a century. Today the average unemployment rate on many western Indian reservations is over 50%, while seasonal unemployment regularly exceeds 70%. The per capita income among western Indians is only 40% of the national average.

Ironically these reservation lands once thought worthless are now known to contain huge quantities of coal and uranium worth inestimable billions of dollars at the marketplace. Of the 50 million acres of Indian land in the West, 13.5 million acres bear more than 80 billion tons worth of recoverable coal reserves. As much as 30% of all western coal and 16% of all coal in the country lie under Indian land.

Such extensive coal reserves—buttressed by Indian ownership of more than half the country's uranium and 3% of its still untapped oil and natural gas reserves—have fueled hopes that energy resource development will lead an impoverished people to prosperity. "Three more years," commented former Crow Tribal Chairman Patrick Stands Over Bull recently, "and if things fall into place our people will at last get a chance to be on top for a change."

The Omnibus Tribal Leasing Act of 1938 allows Indians to lease coal rights to private corporations subject to the regulations and approval of the federal Department of Interior. According to its provisions, six Indian tribes have leased nearly a quarter million acres of their coal land, under 10 separate contracts, to the private sector for coal development.

CEP discovered in its study, *Leased and Lost,* that large tracts of extremely valuable coal-bearing Indian land had been leased to a limited number of large coal mining companies. The Department of Interior—which has the responsibility to set leasing regulations and advise the Indians on important resource management questions—had virtually ignored the interests of the Indian tribes on coal leasing matters. Constantly bargaining from positions of acute poverty, represented by tribal leaders who lacked expertise in the white business world, and ill-advised by the Department of Interior, the Indian tribes repeatedly received far less than fair market value for their coal.

They often granted coal developers rights to scarce water resources vital to their farming economies. Developers were further accorded complete freedom to design, construct and operate any coal mining or conversion facility of their choosing. Except for filling construction jobs of limited duration, developers consistently passed over the Indian labor supply in favor of whites brought in from off the reservation. Lease terms included few if any environmental constraints. Royalty rates averaged only 12¢ for every ton of coal removed from Indian land. Revenue losses to the tribes and environmental degradation to their land were considerable—especially at Utah International's Navajo strip mine, until recently the largest strip mine in the world.

No new leases have been issued since the publication of *Leased and Lost* in 1974, and one has been relinquished. To some degree, then, the recent history of Indian leasing mirrors that of the federal coal leasing program. Leasing activity in both programs was heaviest in the 1960s and early 1970s—80% of the valid Indian leases were issued during this short period.

There are currently 10 coal leases spanning 239,402 acres of Indian land (see Table 3-1). Although they represent only a fraction of the total land under lease in the West, Indian leases have a special importance because of their massive size. The average Indian lease covers 23,940 acres—16 times larger than the average federal and 25 times larger than the average western state lease. Each Indian lease—except a tiny tract on the Uintah and Ouray reservation—is easily large enough to support a massive strip mine operation capable of extracting millions of tons of coal per year. Five large mines are already operating on Indian land in the West, four of which rank among the 10 largest coal mines in the country (see Table 3-2). Output from these five tracts totalled 22.9 million tons in 1977—a level of production which brought the various Indian tribes a total of $6.2 million in royalties. This average royalty rate of roughly 27¢ per ton represents a mile increase over the historical average of 15¢ per ton that CEP found in 1973. The renegotiation of several leases and the appearance of new mining operations—most notably the Absalonka mine on Crow land—explain this increase.

Coal leasing activity on Indian land in the West has concentrated in two regions on lands inhabited by four different tribes. The arid San Juan River Basin lands of the Navajo and Hopi nations in New Mexico and Arizona continue to bear the brunt of Indian coal development. Land controlled by the

TABLE 3-1

INDIAN COAL LEASES

Leaseholder	Acreage	Initial Lease Date	Location
AMAX	14,236	2/15/73	Crow Res., MT
Consolidation Coal Co. and El Paso Natural Gas Co.	40,281	12/2/68	Navajo Res., NM
Gulf Oil Corp.	11,157	9/18/64	Navajo Res., NM
Dr. Roy Humphreys	200	5/12/41	Uintah/Ouray Res., UT
Peabody Coal Co.	24,858	8/28/64	Navajo Res., AZ
Peabody Coal Co.	40,000	{ 6/20/66 7/7/66	Navajo Res., AZ Hopi Res., AZ
Peabody Coal Co.	16,031	12/3/70	Northern Cheyenne Res., MT
Shell Oil Co.	30,247	6/8/72	Crow Res., MT
Utah International, Inc.	31,416	10/21/57	Navajo Res., NM
Westmoreland Resources	30,976	6/14/72	Crow Res., MT

TABLE 3-2

PRODUCING INDIAN COAL LEASES

Leaseholder	Location	Production 1977 (tons coal)	Royalties 1977
Gulf Oil Corp.	Navajo Res., NM	203,712	$40,021
Peabody Coal Co.	Navajo Res., AZ	6,897,578	$1,600,000
Peabody Coal Co.	Navajo and Hopi Res., AZ	4,427,704	$1,815,028
Utah International, Inc.	Navajo Res., NM	6,900,000	$980,000
Westmoreland Resources	Crow Res., MT	4,529,053	$1,800,000

neighboring Crow and Northern Cheyenne tribes in southeastern Montana—with 15 billion tons of strip coal buried underneath—constitute the second locus of activity.

Of the Crow leases, Westmoreland Resources—a consortium of Westmoreland Coal, Morrison & Knudsen, Kewanee Oil, and Penn Virginia—controls the largest and only currently active one. Its 30,976-acre lease tract is the site of the Absalonka mine, the eighth largest coal mine in the country last year. Production here exceeded 4.5 million tons and brought the tribe $1.8 million in royalties in 1977. Shell Oil Co. and AMAX control the other two Crow coal leases, covering 30,247 and 14,236 acres respectively. Neither of these two leases has ever yielded any coal. The Crow have issued an additional five prospecting permits to Peabody Coal Co. and Gulf Mineral Resources. These permits cover 158,413 acres, more than twice the Crow acreage currently under lease. Both permittees have filed lease applications for these tracts.

On the adjoining Northern Cheyenne reservation, Peabody controls a 16,031-acre lease it obtained from the tribe in 1970. No coal has ever been produced on this land, the only Northern Cheyenne land now under lease. The tribe has also issued 11 prospecting permits spanning 227,683 acres—almost the entire reservation—to such energy conglomerates as AMAX, Peabody, Chevron, and Consolidation Coal.

Navajo lands within New Mexico's borders contain three coal leases. Utah International controls a 31,416-acre lease on which the Navajo strip mine —the oldest currently active mine on Indian land in the West—is situated. Production here in 1977 was just shy of seven million tons, making it the country's fourth largest coal mine. The tribe received just under $1 million in royalties based on this production level. Gulf Oil's 11,157-acre lease—the smallest Navajo lease—adjoins federal land on which the McKinley mine had been developed. The portion of the mine situated on Navajo land generated more than 200,000 tons of coal last year. The CONPASO venture of Consolidation Coal and El Paso Natural Gas Co. owns the largest Navajo lease, a 40,281 acre tract that has yet to produce coal. Development on an unprecedented scale could occur here, however, if the two companies go ahead and build mammoth gasification facilities on nearby land.

Navajo lands in Arizona contain an additional two leases—both of which Peabody Coal has brought into production. Situated on 24,858 acres of leased land, the Kayenta mine produced 6,897,578 tons in 1977 and generated $1.6 million in royalties paid to the tribe. It ranked as 1977's fifth largest coal mine. The other lease covers land jointly owned by the Navajo and Hopi tribes—traditional rivals in the Southwest. This 40,000-acre tract is the site of the Black Mesa mine, scene of a bitter struggle between traditional Hopi Indians seeking to protect their land from coal development and the Peabody Coal Co. This, the tenth largest coal mine in the country, produced 4.4 million tons in 1977.

Rounding out the roster of Indian coal leases is a tenth lease on the Unitah and Ouray Reservation in Utah. This tiny 200-acre lease, originally issued in 1941, was purchased in 1975 by Roy Humphreys, a physician from Provo, Utah, from the previous leaseholder, the Red Creek Corp. No production has occurred here since the 1940's. The Uintah/Ouray lease is the only lease to have changed ownership since the original CEP survey in 1973. An eleventh lease in effect in 1973, a 19,452-acre tract owned by Peabody Coal Co. on the Southern Ute Reservation in southwest Colorado, has since been relinquished.

Negotiated in March 1967 and purchased by Peabody for a total payment of $10, this lease quietly expired 10 years later without ever having produced any coal. While the Denver U.S. Geological Survey describes the coal there as "marketable", Peabody showed no interest in renewing the lease there or on any other Southern Ute lands. Currently Sunoco Energy Development Co. is prospecting for coal on the tribe's 307,000-acre reservation. No permit for this exploration activity has been issued however.

THE 1973 REVOLT

The Omnibus Tribal Leasing Act of 1938 requires that Indians, who generally find it difficult to contest the negotiating prowess of the energy conglomerate representatives, be advised by the Department of Interior's Bureau of Indian Affairs (BIA) and United States Geological Survey (USGS) in negotiating coal leases. The resource owning tribes—especially the Crow and Northern Cheyenne—have come to doubt the earnestness of these agencies' efforts on their behalf. The BIA has been perceived as paralyzed by an inherent conflict of interests: on the one hand promoting the best interests of the tribe, and on the other hand, as an agency of the Department of Interior, promoting an increased flow of domestically-derived energy. Such dubious guidance has saddled the tribes with extensive revenue losses and environmental degradation to their land.

In March 1973 the Northern Cheyenne's dissatisfaction with the Department of Interior erupted. The tribe petitioned the Department to cancel all outstanding coal leases and permits on their land. To document their dissatisfaction with leases spanning more than half their reservation, they listed 36 flagrant violations of leasing procedures committed by Interior. These included failure to perform technical pre-leasing examinations, to prepare environmental impact statements, to demand adequate performance and exploration plans, to incorporate environmental protection clauses in the contract, and to make on-site inspections—all required by Departmental regulations. The Cheyenne further implied that Interior sided with prospective lessees rather than the Tribal Councils during the lease negotiations.

On May 19,1974—the day before CEP published *Leased and Lost*—then Secretary of Interior Rogers Morton ruled in favor of the Northern Cheyenne. Agreeing that the violations had occurred, he declared the leases "technically invalid." He pledged to support the tribe's decision to either reinstate the old leases, renegotiate them, reissue new leases to new companies, or cancel all lease arrangements outright.

The Northern Cheyenne are undecided about cancelling or renegotiating the lease and permits on their reservation, however. They have been studying the feasibility of such disparate alternative development schemes as mining their coal reserves themselves, continuing a leasing program for the private sector, and forbidding any coal mining on their lands.

Ever since Secretary Morton's ruling in favor of the Northern Cheyenne, suits, counter-suits, negotiations, renegotiations, Indian council elections and Department of Interior rulings have left the Indian leasing program in disarray. No clear position of the various Indian tribes has emerged in the past four years on the disposition of existing leases. Some leases have been renegotiated, others are being contested in the courts. The future of the leasing and development of Indian coal reserves remains even more uncertain.

The Crow responded almost immediately to the Morton ruling, invalidating a lease with Westmoreland Resources and then signing a new one with that conglomerate in December 1974. The new terms raised the royalty rates from 17.5¢ to 40¢ per ton, at the time, the highest royalty rate in the country. Shortly after, the Crow filed suit in U.S. District Court to have their other two leases, owned by Shell and AMAX, and all prospecting permits cancelled. The lawsuit has dragged on for several years, with the most recent ruling by the U.S. District Court in May 1978 supporting the Crow. The tribe has split into rival factions, each seeking to direct the renegotiating. Despite intense pressure from Interior to reach a settlement, the tribe and the energy companies have yet to agree. Meanwhile the Crow Tribal Council has become increasingly assertive, ordering a tax on developers equal to 25% of the contract value of all coal mined from tribal land.

Other western tribes have sought to renegotiate existing coal leases, but only the Navajo have succeeded. The Navajo lease jointly owned by Consolidation Coal and El Paso Natural Gas was renegotiated in 1977 to include a 12.5% royalty rate. Even this lease had to be renegotiated twice, however, as the Interior Department ruled that the tribe settled for an inadequate royalty rate, 8%, the first time.

The Navajo Nation is now seeking to halt proposed strip mining plans on the Consolidation-El Paso lease, charging the Department of Interior with violating the National Environmental Policy Act, the National Historic Preservation Act, and its trust responsibilities to protect Indian interests. They have further denounced Interior for bypassing the deadline for environmental review of the Four Corners power plant and its toxic emissions. They are charging, in a lawsuit, that the influx of non-Indian personnel on their reservation and the sudden appearance of adjacent boom towns surrounding coal gasification plants would dilute and threaten the tribe's culture.

Even the Uintah and Ouray tribes has lashed out at Interior for allowing the assignment of the 200-acre lease there from the Red Creek Corp. to Dr. Roy Humphreys without tribal concurrance.

A NEW BARGAINING APPROACH

In an effort to gain greater control over the development of their mineral resources, 25 western Indian tribes formed the Council on Energy Resource Tribes (CERT) late in 1975. The formation of CERT signals the abandonment of the long standing suspicion and occasional overt hostility among these tribes as they jointly attempt a profitable development of their resources with minimal destruction of their environment.

CERT was conceived as a bargaining agent for tribes seeking to renegotiate their leases. Under the leadership of Peter MacDonald, the Navajo Chairman, CERT now aims to study Indian resources and goals while providing the tribes with free access to a pool of expertise.

The election of MacDonald, who is not among the most militant of tribal leaders, was contested by many Indian tribes, but even his critics concede that he may be the only one who can effectively lead the new coalition. Already he has hired Ed Gabriel as CERT's executive director. As former head of the Energy Department's Indian Energy Project, Gabriel has strong Washington contacts. MacDonald and Gabriel have staffed a Washington office of 12, and developed plans for a western field office in Denver to eventually employ 15-25 technical experts. They have received $100,000 grants from the Bureau of Indian Affairs and the Economic Development Administration, and are angling for a $700,000 Department of Interior contract to study strip mining regulations on Indian land. The Department of Energy is now evaluating CERT's request for $1.5-$2.5 million to facilitate the establishment of the Denver field office. Though a decision is still pending, Deputy Secretary John O'Leary cheered MacDonald on to "get the best deal you can" at a recent Washington press conference.

In October 1977 CERT talked with Japanese representatives interested in Indian coal, thereby presenting American coal companies with new and unexpected competition. MacDonald has also made overtures to the Organization of Petroleum Exporting Countries (OPEC), requesting advice and even funding from the oil cartel. CERT and OPEC are currently in the midst of a series of talks. MacDonald has urged the coal rich tribes to learn from OPEC's experience that self-sufficiency can be achieved without sacrificing a distinctive culture.

The prospect of an OPEC–Indian alliance has touched off fears that soon the Indians may act as an energy cartel. However, fears of a ruthless Indian cartel frustrating the energy needs of the nation seem premature. Most western Indian tribes do ultimately want some kind of coal development to occur on their lands. They are seeking, however, to halt the kind of development that forces them to absorb the negative impact of coal mining while depriving them of adequate long-range financial, political and social benefits.

RIGHTING THE WRONGS

As the western tribes cooperate to determine the future shape of leasing on their coal land, each tribe in its separate fashion is seeking to resolve the uncertain status of its existing leases.

Northern Cheyenne

Between 1969 and 1971, the Northern Cheyenne auctioned off leases and prospecting permits spanning 43,000 acres—more than half the reservation's total acreage—to such energy corporations as Peabody Coal Co., Consolidation Coal Co., AMAX and Chevron. Bonus payments totalled $2.25 million, an average of $9 an acre. Lease terms called for production royalties ranging from 15¢ to 17.5¢ for each ton of coal mined.

In the summer of 1972, shortly after this divestiture of land, Consolidation Coal Co. presented the Northern Cheyenne Tribal Council with plans for a large strip mine and four gasification plants to be sited on the reservation. To obtain the required 70,000-acre lease tract, Consolidation Coal offered the tribe $35 an acre in land rent, 25¢ per ton in royalty payments, and a bonus of a sorely needed $1.5 million community health center at Lame Deer, the tribal center. Consolidation Coal's offer opened the tribe's eyes to potential benefits from coal leasing. Existing leases and permits were given new scrutiny and on March 5, 1973, instead of accepting Consolidation Coal's offer, the Tribal Council voted unanimously to cancel every outstanding lease. The Cheyenne soon hired the Seattle law firm of Ziontz, Pirtle, Morisset, and Ernstoff—Indian land law experts—to direct the Bureau of Indian Affairs to abrogate all leases and permits.

Then Secretary of Interior Rogers Morton ruled in support of the Northern Cheyenne and in March 1974 encouraged the tribe to renegotiate lease terms. The Northern Cheyenne, however, have yet to officially cancel or renegotiate leases, though they continue to perceive them as invalid. Peabody Coal Co. has continued to make rent and royalty payments into an escrow account pending renegotiation.

It is this pace that distinguishes the Northern Cheyenne from other western tribes such as the Crow. While the Crow reacted to Morton's ruling by immediately seeking to renegotiate its leases for more favorable terms, the Northern Cheyenne have acted more slowly and deliberately, placing a premium on safeguarding their traditions from the impact of coal development. The tribe is now wrestling with the prospect of forbidding any coal development on their lands—at least in the forseeable future. In April 1977 the Northern Cheyenne directed the Environmental Protection Agency (EPA) to redesignate their reservation as a Class I area of pristine air not to be degraded. Under the Clean Air Act amendments of 1977, no industrial expansion is permitted which would increase the air pollution in any area designated as Class I. Indian and state governments can request the EPA Class I status for any part of their lands they choose to protect.

This tribal action was a response to a proposal by a consortium of utilities to expand the existing Colstrip Montana power plant adjacent to the reservation. Already Colstrip units 1 and 2 were emitting pollutants in such quantities, the tribe claimed, that respiratory illnesses were on the increase and agricultural output was suffering. The proposed Colstrip 3 and 4 units would only aggravate these problems. "We're not against progress," explained Northern Cheyenne Tribal Council Chairman Allen Rowland, "but for us, progress means developing our timber and agricultural resources. They are our livelihood and the core of our values. If our air is degraded, these things will be diminished."

The Northern Cheyenne request for Class I designation was the first such exercise of the power granted Indian tribal governments under the Clean Air Act Amendments. By its very suddenness, it has threatened the feasibility of the Colstrip plant. The consortium of utilities behind Colstrip—Montana Power Co., Puget Sound Power & Light, Portland General Electric, Washington Power Co., and Pacific Power & Light—claim they can meet the Class I requirement that power plants' 24-hour ambient concentration not exceed five micrograms per cubic meter. The EPA disagrees, however, and has thrice ruled against the consortium's efforts to exempt the Colstrip expansion from the Class I non-degradation standards. While a U.S. District Court refused to grant an injunction against completion of the plants because they were begun before the Northern Cheyenne requested redesignation, the case is currently on appeal. Thus, the Class I designation holds, delaying Colstrip 3 and 4 another year, at least until the Circuit Court of Appeals announces its decision in late 1978.

The tribe's pro-development neighbors do not share its concept of "progress." The Crow Indians and the state of Wyoming have expressed concern that the protection of Northern Cheyenne pristine air will prevent proposed development on their nearby coal lands. Mining corporations fear that strip mine "fugitive" coal dust may drift onto Northern Cheyenne land and violate the Class I air quality standards there. Wyoming ranchers, Crow Indians, AMAX, Westmoreland Resources, Thermo Resources and the consortium of utilities backing Colstrip, have therefore joined together to protest the tribe's new pristine air designation and ask again for EPA review. As the Northern Cheyenne believe that they should not be forced to accept the environmental degradation resulting from the decisions of their pro-development neighbors, so too the utilities, mining companies and Crow Indians ask not to be penalized by the "go slow" position of the nearby Northern Cheyenne. There is now an amendment to the Clean Air Act before Congress to strip the Indians of the right to request redesignation. Resolution of this controversy may await Congressional decision on this amendment.

Crow

In the mid-nineteenth century, the Crow Indians were rewarded with 58.8 million acres of southeastern Montana's lands, a reservation that has since shrunk to 2.3 million acres. More than three-quarters of the Crow's lands are now owned or leased by non-Indians.

Unlike the Northern Cheyenne, the Crow always have believed that the coal underlying their lands should be developed. In early 1975 the tribe rejected the recommendation of the Crow Coal Research group that a six month moratorium be imposed on the development of reservation coal. In the wake of Morton's ruling invalidating the Northern Cheyenne leases, however, the Crow too have begun to assert themselves. They have cancelled all existing leases and permits for coal development on their lands, but have renegotiated several of them. The Crow still invite coal development, but they now insist that it bring the tribe a fair price.

The only lease to be successfully renegotiated, and thus the only technically valid Crow coal lease, involves the 30,976-acre tract upon which Westmoreland Resources operates its Absalonka strip mine, the eighth largest coal mine in the country. Originally approved in June 1972, it contained a royalty rate of 17.5¢ per ton of coal mined. Immediately after Morton's ruling in May 1974, Westmoreland sought to blunt any tribal reaction by approaching them with an unsolicited offer for an adjustment of lease terms. Westmoreland's offer—raising the royalty payment from 17.5¢ to 25¢ per ton, gradually escalating to 35¢ per ton—was less generous than the terms the Crow had hoped to achieve, however. By the end of June, Crow tribal members voted 225-39 to reject it. Less than one month later, the tribe cancelled the lease.

Westmoreland came right back with a new offer since it needed the leased land to expand its mine to meet long-term contracts. By December 1974 the tribe had approved by an enthusiastic 343-33 vote new lease terms that raised the royalty payment from 17.5¢ to 40¢ per ton, included an escalator clause to ensure that royalties would increase along with the price of coal, required tribal approval before the siting of any gasification plant within 50 miles of the reservation's borders, and guaranteed preferential employment rights for the Crow Indians in Westmoreland mining operations.

Interior recently approved Westmoreland's plans to expand the Absalonka mine by almost 2,000 acres to enable a peak production of more than 10 million tons per year every year from 1981 to 1995. Royalty payments to the tribe are expected to increase from $1.8 million last year to $7 million for each peak production year. And when full employment is reached in 1981, 167 of the 210 mine employees are to be Crow Indians—almost triple the number of Crow employees now.

The two other leases and the prospecting permits on Crow coal land are bogged down in law suits, counter-suits and negotiations. In 1975 the Crow invalidated AMAX's 14,236-acre lease, Shell's 30,247-acre lease, and prospecting permits issued to Gulf Mineral Resources and Peabody Coal Co. spanning 159,413 acres—in short, all coal holdings on the reservation except Westmoreland's. The energy companies are challenging the invalidations in federal court. Since 1974, Shell and AMAX have paid rent and advanced royalties on their lease tracts into an escrow account pending the outcome of litigation.

Two years after the Crow's ruling, there still has been no successful renegotiation of the leases. Factionalism within the Crow tribe itself has helped keep negotiations deadlocked. For the last year, the Crow Coal

Authority has been the tribe's official bargaining agent. Created by tribal resolution and presided over by Tribal Chairman Forrest Horn, the Crow Coal Authority is committed to renegotiating lease contracts to include a profit-sharing royalty structure. Most of its attention has been focused on reaching a compromise with AMAX, which has seemed more willing to accept profit-sharing. At a tribal congress in late 1977, however, loyal supporters of recently-impeached Tribal Chairman, Patrick Stands Over Bull, formed the Crow Coal Council to pursue the aspirations of their old leader. Stands Over Bull is widely known to favor renegotiation with Shell rather than AMAX. Five times the Council has sought to win tribal approval for new negotiations with Shell. Such persistence led to rumors of bribery and corruption, and contributed to the ousting of Stands Over Bull as Tribal Chairman. As Chairman of the Coal Council, he has continued to press for a settlement with Shell to include revised fixed royalties and a sizeable cash bonus payment up front.

In 1978 the Bureau of Indian Affairs proposed the establishment of a new negotiating committee to help heal the rift between the two camps. The Bureau called for each of the Crow reservation's seven districts to elect two representatives to a 14-member negotiating committee. At a recent tribal meeting heavily attended by Crow Coal Authority supporters, the BIA proposal was decisively rejected. The Authority therefore remains the tribe's official coal negotiating voice, while Patrick Stands Over Bull's Coal Council persists as its antagonistic rival.

In the same tribal meeting that rejected the BIA proposal, Crow members voted to explore an unleased coal-bearing tract in the east central portion of their reservation—without the aid of outside energy developers. Bold decisions such as these crop up more frequently now, reflecting in part the new expertise the tribe has hired to help solve its development dilemmas. Charles Lipton, who has advised 22 different governments on the negotiation of natural resource agreements, has been retained as legal counsel. The Department of Interior's Bureau of Mines has spent roughly $100,000 helping the tribe assess the financial implications of varying negotiating positions—the first time such an alliance has ever been struck.

This expert advice has helped the Crow tribe plan more sophisticated scenarios for mineral development. More importantly, perhaps, the Crow now have the added confidence to stand firmly behind their terms—when they can agree on them—even when confronted with proposals that promise the tribe millions of dollars. Already the Crow have rejected a series of proposals as providing inadequate compensation to the tribe.

Shell started the renegotiation process off on a sour note in 1975, insulting the tribe by offering new lease terms that would have included a $200 to $400 bonus payment to each tribe member who, Shell implied, would be paid before the Crow Fair, a festive occasion when tribal families would be facing extra expenses. Shell made a new offer in early 1978 which the tribe is now considering. But sentiment still runs strong against Shell, both for its earlier insulting proposal and for its reluctance to consider anything but a revised fixed royalty.

AMAX, on the other hand, has not only accepted the notion of profit-sharing, but has also accepted in principle a service contract under which the Indians would retain ownership of the coal while the company would mine and market the coal for a fee. Although AMAX has proposed lease terms that would include a royalty rate of 19.3%—more than any other tribe in the nation has received from its coal reserves, and far more than the federal minimum royalty requirement of 12.5%—the tribe is only willing to grant AMAX rights to 220 million tons of coal underlying 4,000 acres. AMAX wants rights to all 450 million tons that lie buried under the 14,000 acres of its original lease tract.

Recently, the Crow have also received proposals from prospectors, Peabody Coal and Gulf Oil. Their prospecting permits have expired, and each company hopes to exercise its option to lease. Gulf has already asked for a lease on most of the 73,000 acres it held under prospecting permit, while Peabody seeks an 11,000-acre coal lease from the tribe. The Crow intend to limit to 2,560 acres the size of any new lease it issues.

Deadlocked in negotiations, Shell, AMAX, Peabody, and Gulf have been ordered by the Interior Department to make substantial progress toward a mining agreement or risk losing their leases completely. The first ultimatum was issued in early October 1977 and gave the companies 60 days to show "sufficient progress in negotiations with the tribe to indicate clearly that the parties will arrive at an agreement in the near future." Shell and AMAX reportedly approached the tribe with new, though still unsatisfactory, offers. The November deadline passed, however, and Interior backed off from any punitive action. In January 1978 Interior gave the coal companies a new 30-day deadline to defend their holdings or risk cancellation. Shell, convinced that it had been earnestly seeking a new agreement with the tribe, labelled this last Interior action an "unwarranted intrusion" and filed suit against Interior in federal district court. Interior has again backed off, deciding to take no further action itself.

Navajo and Hopi

When the United States Commission on Civil Rights turned its attention to the Navajo nation in the first of a series of in-depth analyses of Indian life in America, it reported conditions strikingly similar to those in Third World countries of Latin America, Africa, and Asia—poverty, exploitation and, despite a wealth of natural resources, near bankruptcy. The Navajo birth rate was two and a half times greater than the overall United States rate. Per capita income was less than $1,000 a year. Sixty percent of the households lived without electricity and 80% without water or sewers. Joblessness was pervasive. Alchoholism reached down to afflict even elementary school students.

Such poverty and despair have made the Navajo extremely dependent on the revenues accruing from mineral development on their reservation, which in turn has limited their participation in the western tribes' assault on Interior and the Indian leasing program. Until recently, they have had neither the capital nor the expertise to develop their reserves themselves. Considerable capital requirements make the prospect of joint ventures with established energy developers remote. Unlike the Crow or Northern Cheyenne, the Navajo appear unwilling to risk outright cancellation of their leases. Even renegotiation for better lease terms—which has been their general strategy—has proved difficult for the tribe to pursue vigorously.

The only coal lease successfully renegotiated thus far has been for the 40,281-acre tract issued to Consolidation Coal Co. and El Paso Natural Gas Co. in December 1968. Even this lease had to be renegotiated twice. The first time, Navajo representatives accepted lease terms authorizing a royalty payment of only 8% at a time when the federal government was requiring 12.5% on all federal lands. Interior Secretary Andrus intervened and instructed negotiators to bring the terms into accordance with federal standards. In September 1977 El Paso and Consolidation Coal, fearing outright cancellation of the lease, agreed to new lease terms that included a 12.5% royalty payment, a $5.6 million advance royalty bonus, and a provision calling for a review of the royalty rate in five years.

The Navajo have also sought to bring Utah International to the bargaining table to renegotiate their 31,416-acre lease, the terms of which, according to the tribe's Chief Legal Counsel, George Vlassis, "...could fit on a small post card." In the past, Utah International, which now pays a 15¢ royalty on every ton of coal mined from the Navajo strip mine, had been aloof to the tribe. Now that they need new right-of-way privileges to continue their mining operations, however, Utah International has expressed a willingness to talk. Even so, Vlassis told CEP, the tribe had to exert some muscle by notifying Utah International's customers that unless the lease were renegotiated to the tribe's satisfaction, it would be cancelled. The company then offered to increase the royalty payment to 12.5%. In February 1978 the tribe rejected this proposal. Its representatives are now seeking concessions such as guarantees of employment quotas and management trainee programs.

The Navajo have yet to press Gulf Oil for renegotiation of terms agreed to in September 1964 for a lease covering 11,157 acres adjacent to the energy company's McKinley mine. Development here for the most part has been on federal rather than Indian land; Indian land last year contributed only 203,712 tons to the mine's output of 1.4 million tons. Furthermore, Gulf, unlike most other energy developers on Indian land, has made a genuine effort at upgrading the Navajo work force and maintaining fair labor policies. The tribe has not, therefore, sought to renegotiate the terms of Gulf's lease—which authorize a royalty payment ranging from 25¢ to 37.5¢ per ton of coal mined.

Peabody Coal Co. too has conscientiously sought to meet some of the non-economic aspirations of the Indians from whom it leases land. The labor force at Peabody's Kayenta and Black Mesa strip mines in northeastern Arizona, the latter of which stretches onto land shared by the Navajo and the Hopi, consists of four times more Indians than whites. Still, royalty payments to the Navajo range only from 20¢ to 37.5¢ per ton of coal mined from the Kayenta mine and even less from the Black Mesa mine. Royalties from production at both sites vary with the price and destination of the coal. The Navajo have been unsuccessful in seeking to renegotiate the lease terms. Navajo tribal counsel George Vlassis, blames Interior's aloofness for the deadlocked talks.

Controversy is nothing new to these two Peabody coal leases—they were conceived in it. When Peabody approached the Navajo and Hopi tribes in the mid-sixties to acquire leases, it unveiled plans to strip mine land held sacred by both tribes. The Black Mesa is a land of "shrines and spirits to the Hopi," according to *Audubon* Magazine writer Alvin Josephy, "where man comes close to unity with nature and the supernatural." To the Navajos, the Black Mesa is the Female Mountain, which with its neighboring male counterpart embodies holy harmony. Moreover, much of the land Peabody sought to lease was, and still is, the subject of a longstanding ownership dispute between the Navajo and Hopi that dates back to 1882 when the Hopi Reservation was established in the midst of the more populous, aggressive and expansive Navajos. In both reservations, however, Peabody's proposals to lease almost 65,000 acres pitted traditionalists who resented the prospect of strip mining on nearby sacred land against "progressives" who anticipated much-needed revenues accruing to the tribe.

On both reservations, the pro-development progressives carried the day. In August 1964 the Navajo tribe approved Peabody's request to lease 24,858 acres, now the site of the Kayenta strip mine. Coal mined there now is shipped to the coal-fired Navajo generating station operated by the Salt River Project in Page, Arizona. Less than two years later, the Navajo and Hopi Tribal Councils authorized the leasing of an additional 40,000 acres to Peabody, a tract of land upon which the coal company operates its Black Mesa strip mine. Black Mesa strip coal feeds Southern California Edison's Mohave power plant near Bullhead City, Arizona, which in turn supplies electricity to such mushrooming southwestern metropolitan areas as Southern California, Las Vegas, Phoenix and Tucson.

Neither Peabody coal lease proposal was openly deliberated in the manner intended by the Indian Reorganization Act of 1934. Only six legitimate Hopi Tribal Council members voted to issue Peabody a coal lease. The lease was issued despite the fact that the Council was four members short of a quorum—a flagrant violation of the Hopi constitution. Tribal progressives, it seems, ensured acceptance of the lease proposal by keeping Hopi traditionalists in the dark as to its terms. Josephy reported that five years after the Hopi lease had been issued, "the contract had never been shown or read or explained fully to the Hopi leaders and the people."

Navajo traditionalists were also denied the opportunity to evaluate and debate the merits of Peabody's lease proposals. Kent Smith, a member of the Tribal Council when Peabody's Black Mesa lease was being negotiated, recalled that "we were asked, in effect, to say yes or no to the proposal" with no chance for adequate discussion. This lease too was signed without many Navajo knowing anything about it. Even those Navajo leaders who did approve the issuance of these leases, however, were uninformed of the extent of the coal development and environmental degradation of Peabody's mining in the area. Seventy-five Navajo families lost their grazing lands to Peabody's strip operations. More importantly, the tribe waived its annual rights to 34,100 acre feet of Colorado River water, most of which is now used to cool Salt River Project's power plants. This water, one of the tribe's greatest resources, constitutes two thirds of the Navajo's entire maximum allotment from the river. Some Navajo fear that this diversion could bring ruin to agriculture and ranching on the arid reservation since the tribe is left with only 16,000 acre feet of water a year.

Reluctance to part with any more of the tribe's precious water resources has led the Navajo into their most assertive anti-coal development campaign: to halt mammoth gasification facilities Consolidation Coal Co., El Paso Natural Gas Co., Pacific Lighting Corp., and Texas Eastern Transmission had hoped to construct on the reservation. Until mid-1975, Navajo Tribal Chairman Peter MacDonald had been urging tribal members to support the gasification proposals for the revenue and jobs they would bring to the reservation. His later decision to endorse a moratorium on the diversion of San Juan River water rights from the tribe to the joint CONPASO venture of Consolidation Coal and El Paso Natural Gas for a gasification complex it hoped to construct infused the Navajo challenge to gasification with new energy. On February 1,1978, the Navajo Tribal Council voted to reject a lease proposed by the Western Gasification Company (WESCO), a partnership of Pacific Lighting Corp. of Southern California and Texas Eastern Transmission, to build another gasification complex on the reservation. Within the next 10 days, 13 Navajo tribal members joined the National Indian Youth Council in filing suit to forbid CONPASO from strip mining and gasifying Navajo coal on a 40,281-acre lease on the reservation. The future of both projects is unclear.

Navajo lobbyists have depicted gasification as a costly and inefficient energy technology. They have assailed WESCO's and CONPASO's proposed use of the Lurgi process, an older gasification technology that creates more dust and tar pollution than newer methods. They cite construction cost increases and overruns and the $9 billion total price tag for the proposed plants as evidence of its astronomical expense. The General Accounting Office (GAO) reports, concluding that synthetic gas would be unable to compete with imported oil because of its high cost, are being circulated. The plants, the lobbyists claim, could eventually consume more coal and water than may be available to them. Finally, they cite the unwillingness of private lending institutions to invest in the gasification industry and the companies' subsequent request for multi-billion dollar federal loan guarantees as saddling American taxpayers, in the event of a default, with the absorption of large revenue losses.

Furthermore, the proposed lease terms would shortchange the tribe, contend these staunch antigasification tribal members. The costs would be high and irreversible. Over the course of 25 years, 1.8 billion tons of coal would be strip mined from 58,000 acres of former Navajo grazing land, never, because of its aridity, to be restored to its original condition. Emissions of toxic chemicals such as lead, mercury, and boron would threaten agricultural revitalization stimulated by the new Navajo Indian Irrigation Project. Moreover, this project's scarce water supply may have to be diverted to feed the gasification plants should they require more water than is currently estimated. The Navajo doubt whether the San Juan River can support the gasification plants and the Irrigation Project. Finally, the influx of non-Indians and the appearance of boom towns would, they claim, further disrupt the traditions of tribal life on the reservation.

WESCO's proposal, rejected in early 1978, called for a payment to the tribe of 0.5% of the company's gross revenue—about $2 million—as yearly rent. The Navajo decided that this compensation was inadequate. Even the Department of Interior recommended 3-5% as more equitable, but WESCO claimed that such an increase would price the coal gas out of competition. Only slightly more than half the construction jobs would be filled by Navajo residents, and even these would last only two and one half years. Only 5% of the "permanent" jobs of 25-year tenure would go to tribal members.

WESCO is now developing plans to site its gasification units just beyond the reservation's perimeter. From an off-reservation site, construction and operation of the plants would pose the same threats to the tribe, but offer none of the advantages of siting on the reservation, particularly the power to require preferential employment. WESCO has pushed ahead further and obtained federal approval for rolled-in pricing of its synthetic gas, thereby easing its competitive disadvantage. However, in view of the uncertainties of the synthetics fuel industry, WESCO is not convinced that even the rolled-in pricing agreement will provide sufficient incentive to justify the required $1.3 billion investment. The Navajo, for their part, have launched a lobbying effort to urge President Carter to veto Department of Energy appropriation bill authorizations for a federal loan guarantee program for the synthetics fuel industry. They suspect that the surest way to avoid the impact of gasification plants WESCO and CONPASO seem determined to construct is in the federal government's refusal to subsidize the gasification industry in its infancy.

Buoyed by the preliminary success of their anti-gasification initiative, the Navajo are just beginning to challenge southwestern coal developers. They have consistently rebuked Interior for failing to study the toxic emissions of the Four Corners power plant. Confronted with Interior's unwillingness to investigate the plant, the Navajo have just enacted a new law that fines all power plants operating within the reservation's boundaries for sulfur emissions exceeding one million tons per year. Navajo advisor George Vlassis told CEP that this law could generate as much as $20 million per year for the tribe. However, since the Navajo Tribal Council has previously waived the right to assess taxes on coal developers, Interior is now considering the legality of this new law. Tribal leaders defend it not as a tax, but as a special penalty levied on extraordinary environmental degradation. Whatever Interior's ruling, this issue too should bring the Navajo into the courtroom—where their Northern Cheyenne and Crow neighbors have already been.

Uintah and Ouray

Jackson Moffit, Area Mining Supervisor of Utah's United States Geological Survey told CEP that the 200-acre lease on the Uintah/Ouray Reservation now owned by Dr. Roy Humphreys, is "in and of itself, an unattractive property." Coal seams, as thick as 30 feet, dip steeply there. Underground hydrolic mining seems to be the only possible means of generating a sustained coal yield. The nearest railroad to start the coal on its path to markets is still 150 miles away.

Nevertheless, this oldest and smallest of all Indian coal leases bears the same history of confusion and controversy as the newer mammoth Indian leases on Navajo, Crow and Northern Cheyenne land. In 1941 when the lease was first issued to Heber T. Hall, the mineral rights were owned by the federal government and the surface rights belonged to the U.S. Forest Service. Therefore, the Bureau of Land Management (BLM) first issued the lease. Herber T. Hall immediately began operating an underground mine on the western portion of his federal lease. Local farmers would bring their wagons to Mr. Hall's property every week to buy small quantities of coal for limited home uses. This arrangement lasted only a year, until October 10, 1942, when the mine caught fire and its entrance had to be sealed. Five years later, Hall attempted to strip coal from these old west side workings, but the effort proved unsuccessful. A new underground mine was opened on the lease's eastern half in early 1948.

On April 30, 1948, the exasperated Hall assigned his federal lease to the Red Creek Corp., which held the lease for nearly 30 years. Early in the Red Creek Corp.'s tenure as leaseholder, however, mineral rights on 36,000 acres—including the 200 embraced by this lease—were returned by the BLM to the Uintah/Ouray tribe from which they were earlier ceded. Surface rights, for the time being, were retained by the Forest Service. The lease itself was transferred intact from the Bureau of Land Management to the Bureau of Indian Affairs. The tribe began collecting rental payments. Later, in 1966, the Forest Service transferred the surface, which it had once described as a "valuable part of the National Forest," to the state of Utah, an assignment that angered the Ute Indians.

Otherwise, Red Creek Corp.'s tenure was uneventful there—until May 22,1972, when a representative of the Utah State Division of Wildlife Resources notified the State Branch of Mining Operations that activity at the Red Creek lease site was underway, involving "large amounts of surface disturbance." A quick check with the U.S. Geological Survey indicated that no mining plans had been submitted for authorization, so any activity on the lease was in violation of its terms. Three days later, Charles Albrecht, Mining Engineer for the Branch of Mining Operations, inspected the property and found it "to be worked in a haphazard manner, probably with no overall plan in mind, but rather just to uncover a large quantity of coal for promotional purposes." The environmental impact was considerable. Overburden had been dumped along the banks of the Red Creek. Timber had been cut to catch acid mine water without Bureau of Mines consultations or recommendations. Roads had been cut for drilling and trenching access without the authorization of the county or state.

Eventually, the Branch of Mining Operations, in cooperation with the Bureau of Indian Affairs and U.S. Geological Survey (USGS), traced the operations to Aardvark, Inc., whose representative, Carl Powers, had been seeking access to coal reserves in that area for well over a year. Aardvark Inc., it seems, had been seeking a lease assignment from Red Creek Corp. Before receiving the final assignment, however, and before consulting with the USGS—which must approve all mining plans on federal and Indian lands— Aardvark, Inc. subleased the property to Rio de Oro Mining Co. and instructed it to begin mining operations on the tract. With the approval of Red Creek Corp. and the Bureau of Indian Affairs, Rio de Oro started clearing the area and cutting down trees in preparation for mining. Meanwhile, Aardvark, Inc. released information to a newspaper in the area, the *Uintah Basin Standard* in Roosevelt, Utah, which overestimated the nature and extent of the deposits it hoped to mine. The paper then reported that the coal here would be stripped—despite the fact that because of the steep pitch, stripping had already proved impossible back in the late 1940s; that the reserves were considerable—when in fact, they were unknown; and that the lease spanned 8,000 acres—though it really covered only 200 acres. Aardvark's goal, it seems, was to enlist shareholder support for a mining venture by billing it as a mammoth strip mine overlying a vast acreage of uncommitted reserves. So, Aardvark promoters began bringing prospective shareholders to the scene of the mine to watch Rio de Oro employees removing the overburden to suggest the imminence of large-scale strip mining.

Rio de Oro's preliminary operations on this tract not only violated the terms of the lease but also a number of state mining regulations. Laws regarding timber removal, water pollution, changing the character of a stream bed, and diverting water from its normal course all were broken. In addition, the state charged the mining company with causing irreparable damage to the land surface. The USGS ordered that the mine be shut down on June 1, 1972. It wasn't: when Area Mining Supervisor Moffit inspected the lease area one week later, mining was still in progress. He again ordered that all work be stopped. This time, Aardvark listened. They absconded with some coal and left the area, without every paying Rio de Oro's employees. Those checks the company did write bounced. The Uintah/Ouray never received a penny in royalties.

The state, however, brought Aardvark to court for muddying the waters that fed into a nearby reservoir. Meanwhile, the Securities and Exchange Commission (SEC) had been alerted to the fraud. It too prosecuted and won conviction of the masterminds of Aardvark's stockpeddling scheme for the illegal promotion and selling of stock.

Things have since quieted down on this 200-acre lease tract, but not by much. On September 16, 1975, Red Creek Corp. assigned the lease to Dr. Roy W. Humphreys of Provo, Utah. The Uintah/Ouray, however, question the legitimacy of the assignment since it occurred without their consent. Apparently, the Bureau of Indian Affairs approved the lease assignment without ever consulting or obtaining the necessary concurrence of the tribe. Furthermore, the BIA never advised the tribe that, with a new prospective lessee at hand, the time was ripe for renegotiation of lease terms that had not been adjusted since the lease was first issued in 1941. The Uintah/Ouray, for their part, have fused their two grievances—invalid lease assignment and outdated lease terms—into a coherent position: only renegotiation of the lease terms can legalize the assignment. Otherwise, they continue to view Humphrey's lease as invalid. Little progress has been made to date at the renegotiation table.

CONCLUSION

Despite the hot controversy surrounding coal development projects on the 10 outstanding Indian leases, Indian coal could emerge at any time as the focal point of western coal development. In the past, extensive regulatory requirements and the difficulties of dealing with the tribes directly served as disincentives to development. But this situation is changing. The nature and timing of the new federal coal leasing program—long mired in a moratorium—remain unclear. State-owned parcels generally occur in small tracts incapable of sustaining mining operations. Some state governments are moving to toughen their lax leasing regulations. Thus, the millions of tons of uncommitted strippable reserves buried under massive tracts of Indian land are beginning to appear more attractive to developers.

The Indian tribes in the West are showing an increasing sophistication on mineral development questions in the 1970s which could make them a more formidable, though still desirable, participant at the negotiation table. Most tribes are receptive to offers put forth by energy developers for leases or other contractual arrangements facilitating development of coal reserves. The tribes acting together in the Council on Energy Resource Tribes provide an additional point of access for energy developers.

A recent Department of Interior ruling suggesting that Indian coal may be free from state taxation could supply another impetus to corporations seeking Indian leases. Interior's ruling concerns only $55,000 Montana sought to collect in 1975 under a since-repealed property tax on royalties. State officials, however, suspect the ruling could be broadened to prevent collection of the Montana severence tax which now averages 22.5% of the value of the coal mined in the state. Since other western states have also adopted stiffer severance taxes in the last few years, Indian coal—possible free from severance taxations—might become more attractive yet.

The Indian tribes, however, have by no means decided on a course of action concerning resource development. Even the status of existing leases and mining operations could change momentarily with a Tribal Council vote on any of the five reservations now sustaining coal development activity.

Leasing Indian coal lands thus presents corporate energy developers with both advantages and disadvantages. So, too, for the tribes themselves. There is much to encourage the Indians to lease their lands. Federal law permits them ample freedom to create leasing arrangements of their own accord. Many see the promotion of resource development on reservation lands as an escape route from the devastating poverty that afflicts every western tribe. Equally strong are the fears that coal development would destroy existing Indian culture and institutions. Those opposed to coal development see it as a one-time harvest in which developers operating under present contracts deny the tribes future, far greater returns. Thus, many western Indians believe in "no growth," "slow growth," or "growth only under the supervision of the tribes themselves" as the only alternatives.

Whatever the outcome, the Indian tribes are a force to be reckoned with, not only by energy developers seeking new commercial opportunities, but by the American people as a whole seeking assurances that a coherent national energy policy will be implemented.

STATE PROFILES

4. COLORADO

INTRODUCTION

Colorado's rugged terrain presented many early American pioneers with their first encounter with the West. Believing the vast network of Rocky Mountains, fertile valleys and grassy plateaus foretold a prosperous agricultural economy, they settled there. Prospectors and grubstake miners looked at the towering Rockies and anticipated treasures of precious gems buried beneath the rock. When gold was discovered in 1858, their dreams came true, and their numbers multiplied. Still others, their determination wearied by crossing the pitiless Rockies, remained in Colorado to avoid the further rigors of the unknown.

Large scale settlement, therefore, occurred somewhat earlier in Colorado than in other western territories; lands were staked out, homes were built, and loosely based communities were established. Ranching, agriculture, mining, industry and commerce evolved in the mid-nineteenth century. Later in the century, coal mined from the deep underground reserves crisscrossing the state became the energy resource most commonly and reliably used to heat homes and fuel industries.

The diverse and scattered settlement patterns of Colorado's early inhabitants are reflected in the state's decentralized coal mining industry. Since mining began in 1864, a profusion of mostly small, deep mines has contributed to a cumulative statewide production of 618 million tons of coal, more than any other state. Forty-three mines are currently producing coal, again more than in any other state. These mines, most of which employ underground mining, had a combined output of only 12 million tons in 1977, however. The massive strip mining operations slated for Wyoming and Montana suggest that they may soon displace Colorado as the historical leader in western coal production.

The currently active mines are distributed across federal, state, county and private land tracts which pepper the Colorado landscape. Federal lands have yielded 38 million tons of coal since 1926, the third highest total in the West. More coal has been produced on state land in Colorado than in any other western state. Even so, private land has traditionally accounted for most of Colorado's output. Thirty-four of the 43 currently producing mines are situated solely on private land. Such extensive activity on state and private land is unparalleled in the West.

Colorado's total coal reserves have been conservatively estimated at 14.8 billion tons. CEP interviewed some state officials in Colorado who believed that this figure considerably underestimated the state's mineral wealth. Still, Colorado cannot match the massive reserves of Wyoming or Montana, though it exceeds those of Utah and New Mexico. Most of Colorado's coal is of a high quality bituminous grade, accounting for 46.3% of all bituminous coal reserves in the West. Like Utah's high quality reserves, nearly all of Colorado's coal lies deeply underground, accessible only by underground mining techniques. Only a small fraction of these bituminous coal reserves are close enough to the surface to be strip mined. Nevertheless, these coal deposits have attracted the attention of developers over the last decade, and they now account for a large share of mining activity.

Coal lands in Colorado span nearly 30,000 square miles and include eight different coal regions. Colorado's major coal-producing region lies in the Green River coal fields in the northwestern corner of the state. It is here that most of the state's strippable, high-heating quality bituminous coal is concentrated. The federal government is the dominant land owner in this area, controlling about 82% of 1.8 million acres of coal-bearing land. Thirty-one federal coal leases have been issued for coal development in the two counties, Moffat and Routt, which cover most of this coal field. Four of the 14 active federal coal leases in Colorado lie in the Green River Coal field.

South of the Green River fields lies the Uinta coal field. The region's extensive reserves once attracted considerable mining activity, but uncertainty of water supply has recently discouraged large-scale development plans. Market potential for this coal appears very limited. The construction of a nearby gasification plant, now in the preliminary planning stage, could change that.

The rugged lands of the San Juan River coal region, just south of Uinta, include several small private mines operating to supply local needs. The mountainous Raton Mesa field in southcentral Colorado contains half a million acres of coal-bearing land, mostly in Los Animas County. The federal government controls about one-third of these lands. Production so far has been on private land. In the flat plains of eastern Colorado, the Denver Basin and Canon City fields bear some coal-producing potential. Though strippable, this coal is often of lower quality than that found in the western part of the state. Nevertheless plans for several large strip mines in these fields have been announced in recent years, because they are close to the population and industrial centers of the state.

In the future, small underground mines will still play a large, though declining, role. By 1977, two-thirds of Colorado's coal output emerged from strip mines. A few more large strip operations, similar to those slated for Wyoming and Montana, are already on the drawing boards of energy companies. Complex land ownership patterns will continue to favor the historically-evolved decentralization of the state's coal industry, however.

FEDERAL LEASING

The federal government controls 53% of Colorado's coal-bearing acreage. Most of the 8.7 million acres of federal coal land lies within the Green River coal field in the state's northwestern corner where most coal development occurs.

Leasing of federal lands in Colorado has been extensive compared to most other western states. Only Utah has more outstanding federal coal leases than Colorado's 113; only Utah and Wyoming boast more federal coal reserves under lease than Colorado's 1.6 billion tons. More federal acreage has been leased in Colorado—121,417 acres—than in Montana, New Mexico, and North Dakota combined.

Little leasing activity has occurred in Colorado since 1973. One new preference right lease and one new competitive lease have been issued. One large lease has been relinquished, causing a decline in Colorado's federal acreage under lease. Colorado is one of only two western states to experience this decline—the other being North Dakota.

This slow pace of federal leasing activity may be reversed, however. More preference right lease applications—41—for federal land are pending in Colorado than in any western state except Wyoming. These applications cover more than 93,000 acres and one billion tons of reserves.

Coal mining on federal land in Colorado increased throughout the 1970s although the rate of increase has been topped by several other western states. In 1977, 14 active federal leases generated 3,890,829 tons of coal. This 27% increase over federal production in 1973 represents the second lowest production increase in the West. The Bureau of Land Management forecasts that coal development on Colorado's federal land will increase steadily throughout the century at a rate swifter than that forecast for the West at large but slower than the tenfold increases expected in North Dakota and Utah.

Colorado coal development may receive an unexpected boost due to the random mingling of federal, state and private land across the landscape. Since checkerboarding of these tracts is more extensive in Colorado than anywhere else in the West, developers, until recently, have looked elsewhere to site their mammoth mines and plants. These developers now see more opportunity in Colorado than any other state to capitalize upon a loophole in the short-term federal coal leasing criteria. The Department of Interior's short-term program allows companies mining on private or state land to lease adjoining federal land if they need the federal reserves to sustain mining operations. Extensive checkerboarding in Colorado provides many opportunities to enact this scheme.

Corporate awareness of this loophole surfaced in mid-1977 when Secretary of Interior Andrus authorized the sale of 2,230 acres in Paonia, Colorado to Colorado Westmoreland, Inc. Colorado Westmoreland claimed that the federal land was vital to the continued operation of its Orchard Valley mine, situated on an adjoining 120-acre tract of private land the company had purchased just a year before. The company invested $40 million in developing this land—an astronomical sum for such a tiny tract. Colorado Westmoreland also negotiated a contract with Northern Indiana Public Service Co., calling for production and delivery of 700,000 tons per year. Later, Colorado Westmoreland signed a second contract with the same utility, obligating the coal developer to deliver an additional 500,000 tons each year.

Neither contract could be fulfilled without coal from adjoining federal land. So Colorado Westmoreland applied for a federal lease, claiming compliance with the short-term criteria. Local citizens complained that if Interior issued the lease, other energy companies would soon move into Colorado under the same loophole. Environmentalists added that no Environmental Impact Statement had been completed, so the lease could not be issued. Mark Welsh of the Western Slope Energy Research Center reported that Colorado Westmoreland had actually pre-dated its second contract with the Indiana utility from November to September 1977, five days *before* Judge Pratt established stricter short-term leasing guidelines.

Interior authorized the lease sale over these protests, ruling that Colorado Westmoreland was entitled to fulfill the terms of the first contract, the one signed prior to the lease application. The company was not permitted to lease additional coal land to fulfull the second contract, however. In late February 1978, Colorado Westmoreland was the successful bidder for the lease tract, offering $100 per acre. The only other offer was a token protest bid submitted by the Rotton Apple Coal Co.

Westmoreland's success in getting the lease reflects to some degree the persuasiveness of its argument that it always had intended to expand the Orchard Valley mine. Still, the company's strategy—buying and mining small tracts of private land adjacent to unleased federal land, signing extensive long-term contracts, and then applying for a federal lease to meet the contracts can be easily replicated in Colorado. In this vein, Satchiko Nakazono of the Denver Bureau of Land Management told CEP that more requests for new federal coal leases are pending in Colorado than any other western state.

STATE LEASING

"We have some real kickers in our new leases," exclaimed Tom Bretz, Mineral Director of the Colorado Board of Land Commissioners. CEP agrees—with reservations. In some ways, Colorado's state leasing program is now among the strongest in the West. For example, the state has promulgated new regulations which may retard the speculative holding of land under lease. In other ways, however, the new leasing regulations are confused and contradictory, closing some loopholes while leaving others gaping. On the whole, the state's efforts to manage its coal resources are not as effective as Montana's or North Dakota's.

State coal leasing dates back further in Colorado than anywhere else in the West. The archives of the Colorado Board of Land Commissioners contain records of leases issued as long ago as 1900. Coal was first mined from state land in 1908. Production peaked at 1.9 million tons per year in 1926. State land in Colorado has produced more coal—21.3 million tons in all—than anywhere else in the West.

The modern day Colorado state coal leasing program began in 1954 when Utah International obtained rights to a 9,721-acre tract. Leases issued earlier have since expired. Only three leases were issued between 1954 and 1960. Nineteen others were issued in the 1960s. The Colorado leasing boom started in 1971 when the Interior Department declared a moratorium on the issuance of federal coal leases. Large numbers of energy companies and land speculators then turned toward state land for additional coal reserves (see Table 4-1).

TABLE 4-1

COLORADO STATE LEASE ISSUANCE

Years	Number of Leases Issued
1954-1958	3
1959-1963	4
1964	4
1965	1
1966	0
1967	1
1968	6
1969	3
1970	12
1971	8
1972	15
1973	10
1974	20
1975	19
1976	2
1977	41

TABLE 4-2

REVENUE FROM COLORADO STATE LEASE SALES

Method of Leasing	Number of Leases	Total Revenue Collected
Open Land	91	$21,468
Segregation	9	90
Competitive Bidding	47	102,549
TOTAL	147	$124,107

The Colorado Board of Land Commissioners has issued 147 currently valid leases covering 252,199 acres. Its state coal acreage under lease ranks third among the western states. Roughly three million acres constituting 91% of state-owned coal land remains unleased.

Sixty-nine per cent of Colorado's leased coal tracts are smaller than 1,000 acres, reflecting the scattered, non-contiguous parcels granted the state upon its admission to the Union. There are, however, four massive lease tracts, each larger than 10,000 acres. Nowhere else in the West can such large state land blocs be found. Colorado obtained them when the federal government allowed the new state to exchange small sections of land in the Rocky Mountains or in federally-controlled national forests for larger continuous tracts in other regions. The state fortuitously selected large tracts of land in coal-bearing regions.

Colorado's generosity in leasing its large tracts of coal land is astounding. Most western states prohibit the leasing of tracts larger than 2,500 acres. The Mineral Leasing Act governing the federal coal leasing program limits the size of leases to 2,560 acres. Colorado law contains no such acreage limitation. It is the only western state to have leased tracts larger than 10,000 acres.

Each of these massive tracts was virtually given away under open land leasing procedures—one of a variety of lease issuance procedures employed by the Colorado Board of Land Commissioners. The different procedures are linked, however, by a consistent failure to generate significant revenues to the state (see Table 4-2). Colorado has received an average of just $844 for issuing each of the 147 leases.

Open land leasing, permitting any person or corporation to obtain a lease by paying a filing fee of $10, was employed exclusively until 1975. CEP found 91 leases, including the four largest representing 205,585 acres, or 62% of all land leased to date, issued according to open land provisions.

Another nine leases covering 13,640 acres were created by segregation from existing leases. The state charges a filing fee of $10 for segregating a lease.

In 1975 and again in 1977, the state awarded leases after oral competitive bidding at open auction. Twelve leases were issued at an auction held on June 18, 1975. Another 35 leases were issued at two modified competitive lease sales in July 1977. These 47 competitive leases cover 32,974 acres of land, auctioned for an average of $3.11 an acre—the third lowest among the five states employing competitive leasing.

The Board of Land Commissioners does not keep records of the number of bidders appearing at its lease auctions, but CEP found that 13 of the 47 leases were issued without payment of any cash bonus, signifying that only one party was interested in them.

CEP examined records of Colorado's two competitive leasing programs to better understand the state's inability to generate significant revenues from the competitive issuance of coal leases. Each of the 12 leases issued at the June 1975 competitive lease sale was selected from a list of industry-nominated tracts. The state made no effort to determine or promote competitive interest in them before the lease sale. At the public auction, five of the 12 drew only the minimum opening bid of 25¢ per acre. The state, according to Mineral Director Bretz, was shaken by these meager results. Its first reaction was to throw in the towel and revert to the open land system for the issuance of 11 leases. Then for a 10-month period in 1976, the state maintained a moratorium on all leasing and used the time to rethink its leasing program and develop new guidelines for competitive leasing.

The new competitive leasing program was unveiled in 1977. Its provisions permit companies to nominate land tracts for leasing to the Board. The Board is then to notify roughly 20 relevant state, county and federal agencies of the nomination. These agencies have six months to report objections to the proposed lease tract. The Board weighs these objections and ultimately decides whether or not to lease the land. If it decides to lease, the Board notifies its select mailing list of the tract's availability. At the auction, any company willing to pay the first year's rent ($1 per acre) qualifies for the lease unless another bidder offers an additional cash bonus payment.

This new program was tested at two competitive lease sales in July 1977. The results were dismal. Twelve of the 32 leases issued at these two auctions required no cash bonus payment because there were no bidders to force a cash offer. The highest bid for any lease was $18 per acre. Again, the Board of Land Commissioners failed to regulate the number, size or rate of lease issuance. Only industry-nominated tracts were offered, though the state is empowered to offer tracts of its own choosing. The Board never exercised its power to refuse to offer an industry-nominated tract for lease. Nor did it reject any bid for failing to reflect the fair market value of the land as determined by the state.

Entry into the Colorado state leasing program is further facilitated by the assignment process. CEP found that the current owners of 48 leases, constituting 33% of all valid leases, obtained their holdings on the assignment market. Assignment activity peaked in 1976, when the state maintained a 10-month moratorium on the issuance of new leases: 15 were bartered that year.

Four of the top five state leaseholders in Colorado obtained all their holdings on the assignment market. Only Consolidation Coal, the third largest was the original lessee of any of the state land it now controls. It leased all five of its tracts directly from the state, paying a total of $2,150 for the open land rights to more than 21,000 acres.

The top five leaseholders control 44% of the land under lease, an average level of concentration in the West (see Table 4-3). These same five control only 12% of the leases, however, the lowest concentration of lease holdings among the six western states. The average leaseholder controls only 2.3 leases, also the lowest figure in the western states. Complex land ownership patterns, the effect of the mountainous terrain in scattering coal developers and isolating coal development projects, and the long history of coal mining in the state have enabled a larger and more diversified collection of mining interests to participate in the state leasing program in Colorado than anywhere else in the West.

Of the 63 Colorado state coal lessees, only three produced any coal in 1977, and only five have ever produced. Empire Energy Co.'s Wise Hill mine generated 24,526 tons of state coal in 1977, bringing cumulative production on the state tract there up to 1.1 million tons. Energy Fuels extracted 104,284 tons from its state tract last year. Morgan Coal Co.'s Seneca no. 2 mine opened on state land in May 1977 and is now producing roughly 100,000 tons each month.

State production in 1977 totalled 218,006 tons and brought the state $119,279 in royalties. This average royalty rate of 55¢ per ton ranks as the highest in the West.

Some holders of inactive leases are required to pay advanced royalties, the rates varying with the age of the lease. Leases issued before 1975 charge small advanced royalty rates, if any at all, payable in one lump sum at the beginning of the sixth year of the lease and every year thereafter. Leases issued since 1975 assess stiffer rates. Advanced royalties for these leases begin at $6 per acre per year in the sixth year of the lease and rise to $10 per acre per year in the tenth year of the lease. Advanced royalty collections in 1977 totalled $55,160.

Lessees are also assessed rental rates which vary from 25¢ to $1 per acre

TABLE 4-3

TOP FIVE COLORADO STATE LEASEHOLDERS

Leaseholder	Number of Leases	Acreage under Lease	% Total Acreage under Lease
Mapco, Inc.	2	26,459	10.5
Franklin Real Estate	4	25,312	10.0
Consolidation Coal Co.	5	21,621	8.6
Mobil	4	20,904	8.3
Limon Fuels	3	16,800	6.7
TOTAL	18	111,096	44.0

TABLE 4-4

BENEFICIARIES OF COLORADO STATE LANDS

Institution	Fund Balance
Public School	$58,413,052.66
Penetentiary	20,028.00
Public Buildings	93,624.86
Internal Improvements	622,946.95
Saline	37,757.93
Colorado State University	116,685.54
Hesperus	15,731.60
University of Colorado	44,421.82
Colorado State Forest	40,333.02
TOTAL	$59,404,582.47

per year. In 1977, rental payments accounted for $145,569—exceeding royalties and advanced royalties.

Rental, royalty and advanced royalty collections totalled $320,008 in 1977, representing only 4% of the revenue generated from all minerals activity on state-owned land.

Eight institutions are the beneficiaries of coal development on Colorado land (see Table 4-4). The vast majority of state land in Colorado—as in other western states—has been assigned to the public schools and state universities. The remainder is scattered among a few small institutions and funds.

Rental payments are turned over to the various institutions to become part of their operating expenses for the coming year. Advanced royalty and royalty payments are deposited in the state's Permanent Fund. This Fund is then invested, and its interest is divided among the nine institutions annually. Each institution's share of the interest is determined by the proportion of its contribution to the principal of the fund over the years. Interest income, like rental payments, may be spent by the institutions. The fund's principal, never spent, grows annually. Its current balance is $59.4 million, the lowest in the West. (See Table 4-4).

New leasing regulations promulgated in 1976, if properly enforced, may change the complexion of the Colorado state coal leasing program. Royalty rates, which were as low as 15¢ per ton, were raised to 12.5% of the gross value of strip coal and 8% of the value of deep coal—the highest royalty rates charged by any state leasing program. Advanced royalty rates were reevaluated to reach $10 per acre per year by the tenth year of the lease. Lease terms were shortened to expire after 10 rather than 20 years. The new regulations still grant lessees the preferential right to renew leases, but only if they demonstrate that their land is included in a mining plan or logical mining unit which will produce coal within 10 years, *and* if they continue to pay $10 per acre per year in advanced royalties.

The new regulations attempt to control lease speculation by limiting lease assignments. The state is now empowered to cancel any coal lease assigned with an overriding royalty exceeding 2% of the gross value of the coal.

There is one final "kicker" in the new leasing code. The federal government and all other western state governments have so far denied themselves the discretionary power even to ask the amount of cash bonus payments accompanying lease assignments, but in Colorado, lessees must not only report the amount but also pay the state 10% of the cash payment.

This provision is certainly two-edged. On the one hand, it discourages speculation by increasing the cost of assignment and requiring speculators to disclose their transactions. On the other hand, it gives the state a vested interest in assignments and makes it a beneficiary of lease speculation. The more assignments, the more money the state receives. It remains unclear how the Board will balance its dual responsibilities: to regulate coal development—which might be best fulfilled by stopping speculation altogether, and to raise revenues for the eight institutions it serves—which might be best accomplished by fostering lease assignments. If its past performance in coordinating lease sales to generate competitive interest and regulating development on state coal land is an accurate gauge, the Board will probably sit back and cash the checks it receives from the assignments of lease speculators. In this light, Bretz' defense of the new assignment regulation is illuminating. He told CEP that "speculators do a fine job of putting together a development package."

DEVELOPMENT

Future coal development in Colorado will be dominated by expansion and reopening of underground mines in the western part of the state, complemented by a few new mammoth strip mines similar to those forecast for other western states. Six once-active mines are scheduled to reopen and achieve modest production levels. Each will again employ underground mining methods. Operations in 31 currently active mines will be expanded over the next decade; two-thirds of them will produce under one million tons per year. Small Colorado-based mining companies such as Midland Coal Co., American Fuels, Inc., Energy and Export Co. Ltd., and Eagle Head Coal Co. will continue to mine Colorado's underground seams.

Massive coal operations also have their place in Colorado's coal future. Energy Fuels Corp. plans on expanding its Routt County strip operations—currently the largest in the state—from 2.5 to 4.5 million tons per year by 1980. Two Colowyo ventures jointly-owned by W. R. Grace and Co. and Hanna Mining Co. each promise 3.1 million tons per year. Utah International's new strip mine in Moffat County is expected to produce almost three million tons per year by 1980. Merchers Petroleum Co., T. C. Woodward, and Consolidation Coal Co. own adjoining tracts of strippable coal land in Routt County expected to yield 4.4 million tons per year by 1980. Cameron Engineers of Denver has announced its intention to operate a strip mine in Adams County in the eastern part of the state that would generate 10 million tons per year to feed a nearby gasification plant.

Even without gasification, much of Colorado's coal production will be marketed within its own region. The Colowyo coal ventures will supply power plants in Colorado Springs. Adolph Coors Co. will mine and burn coal in industries it owns within the state. Energy Fuels will continue to ship some coal to power plants owned by Public Service Co. of Colorado.

Increasing amounts of Colorado coal will be exported from the state, however, mainly in an easterly direction. Colorado Westmoreland, Inc. will soon begin supplying Northern Indian Public Service Co. Colowyo's ventures will supply coal for industrial uses in Texas. The American Electric Power system will receive coal shipped from its captive mines in Colorado. The Tipperary Oil and Gas Corp. will sell roughly half a million tons a year to Western Kansas Iron Ore reduction plants starting in 1979. Recently, CF&I Steel Corp. has emerged as the leading wholesaler between several small Colorado mining operations and large eastern and midwestern markets. A longtime customer of western railroads and transporters, CF&I is currently seeking markets for coal mined by Western Carbon Co., Western Slope Carbon Co., and Empire Energy Corp., as well as for coal mined from its own operation in Las Animas County.

STATE POSTURE

No matter what the pace of Colorado coal development, constant scrutiny from a myriad of newly-created state and community agencies is assured. An ambitious bureaucratic network of organizations seeks to monitor coal development and alleviate its impact. Despite the unparalleled energy and enthusiasm of agency staffs, CEP is less than sanguine about their capacity to effectively contain coal development's impact. In the words of Joan Martin, Colorado's Coal Coordinator, many of the new state agencies exist "as a layer of liberal frosting" concealing a nonchalant pro-development posture. Others echoed her complaint that Colorado's coal bureaucracy succeeds more in accommodating than regulating development.

The widespread concern in Colorado over coal development sprang partly from an earlier fear of an oil shale boom. In the wake of the Arab oil embargo, the federal government in 1974 accelerated the implementation of its oil shale development plan, the cornerstone of which was the leasing of four tracts of shale-rich land—two in Colorado and two in Utah. Each 5,000 acres, the two Colorado tracts contained the richest shale seams in the country and were expected to be the sites of the nation's first commercial oil shale operations. When the federal government auctioned off the two Colorado lease tracts, the winning bidders—all oil companies, including Gulf and Shell—bid an incredible $328.1 million for the rights to develop oil shale reserves. This was more revenue than Interior garnered from the issuance of all federal coal leases in all western states over the entire history of the federal coal leasing program.

Colorado received $73.8 million as its share of the cash bonuses. Some of these funds were used to establish the Oil Shale Coordinators Office to assist local, state and federal authorities in oil shale planning and coordination. Most of the money was deposited in the Oil Shale Trust Fund where it remains with a balance exceeding $70 million at the end of 1977.

Large scale oil shale activity has yet to materialize, however. Rather than eliminating the Oil Shale Coordinators Office, the state renamed it the Socio-Economic Impact Coordinators Office (SEICO) and instructed it to monitor all types of energy development, including coal. By 1977, the state had also created the Department of Local Affairs, the Energy Impact Assistance Advisory Committee, the Energy Policy Council, the Western Slope Energy Impact Advisory Council, the Mine Land Reclamation Board, and the Office of Rural Development, to work with the long-standing Department of Natural Resources on coal development issues. The precise relationships among these organizations are complex, perhaps indefinable, permeated by random overlapping, unforeseen voids and lofty intentions.

The Socio-Economic Impact Coordinators Office intersects these and other agencies. Among its numerous functions are monitoring statewide energy activity, identifying communities affected by development, creating and developing local impact assistance teams, compiling data on affected communities, and developing a boom-town financing study. Along with the Energy Impact Advisory Committee of the Department of Local Affairs and the Office of Rural Development, SEICO assists local communities in the disbursal of federal and state funds.

That these agencies have sprung from, and in turn created, widespread concern over Colorado coal development seems incontestable. But while CEP found activity occurring at a frenetic pace, much of the work within these agencies seemed nonproductive. John Fernandez of SEICO talked to CEP in December 1977 in his overcrowded office. Several times within the last few months of 1977, Fernandez's office had been moved, preventing him, he complained, from sustaining substantial projects. Such frequent transplants have left his office in shambles. Boxes and crates served as file cabinets. Unfinished partitions divided off miniscule work spaces. Workers, though enthusiastic, frequently interrupted to ask the whereabouts of important documents. The noise level was astounding. From these chaotic environs, Fernandez confessed that "the state is not very far along in planning to meet the impacts of energy development." He further noted that a great deal of energy was being spent to satisfy political motives rather than to meet the genuine needs of affected areas. Others were more strenuous in faulting SEICO. Carolyn Johnson of the Environmental Policy Institute told CEP: "I have not seen one single case where

SEICO has made any difference. Rather than control impacts, they just pay for them."

But even the most efficient bureaucratic network cannot produce results in the absence of adequate revenues, and Colorado has been less than energetic in its pursuit of coal-related taxes. Until January 1978 the only tax on coal mined in Colorado was a mine levy of 0.7¢ per ton. Though the tax was established to subsidize the cost of mine inspections, the revenues were diverted to a State General Fund never to be expended on coal mining maintenance or inspections. Norman Blake, Director of the Division of Mines, told CEP that he never received these funds, earmarked for his office, without a fight from the legislature.

Until January 1, 1978, when the state's first severance tax became effective, Colorado was one of only two western states—the other is Utah—not to levy a tax on the privilege of extracting coal. The new tax is among the lowest in the West. Strip mine operators will be required to pay 60¢ per ton in severance taxes to the state. Underground mine operators pay 30¢ per ton. Both strip and underground miners are permitted to exempt the first 8,000 tons of coal mined each quarter of the calendar year. Such provisions distinguish the small miner from the mammoth strip miner, but they also generate less revenue. George Tung of the Revenue Department's Office of Planning and Budget defended this tax structure, telling CEP that "we don't want to chase industry away in Colorado."

Recently, the Colorado legislature revised the formula for allocation of severance tax revenues to better meet the special needs of affected areas. The General Fund's share of severance tax revenue will decrease from 40% this fiscal year to 20% in Fiscal 1981. The Local Government Severance Tax Fund, created specifically by the legislation authorizing the tax, will eventually receive 45% of the collected revenues. These funds will be used to help provide and maintain public facilities in affected counties and municipalities. The third recipient of severance tax revenues is the newly-created State Severance Tax Trust Fund, whose share of collected revenue will rise from 15% in Fiscal 1978 to 35% in Fiscal 1981. However, the precise nature of the expenditures from this fund have yet to be determined.

The Colorado legislature has also promulgated new guidelines for the allocation of revenues accruing to the state from federal coal leasing within its boundaries. The Federal Leasing Act Amendments of 1975 accord each state 50% of all revenues derived from federal coal leasing within the state. Colorado now grants half of its share to the county where the revenue was derived, 25% to the state-wide public school fund, 15% to the Local Government Mineral Impact Fund, and 10% to the Colorado Water Conservation Board. The legislature established a ceiling of $200,000 on the revenue that can be directed towards any single county. Any revenues beyond $200,000 spill over into the public school fund. Already Routt and Moffat Counties, the foci of Colorado coal activity, have lost substantial revenues to the school fund because of this ceiling. Fernandez of SEICO admitted that "the schools and not the counties are the big winners in the federal leasing game here.''

Pressure from environmentalists to increase and better distribute revenues from coal development to state coffers and alleviate adverse coal impacts have been sporadic and ineffective. Joan Martin told CEP that environmental organizations within the state, while numerous, are ''poor lobbyists.'' She suggested that dissension from within has weakened the Colorado environmental movement and reduced its persuasiveness at the legislative level. Environmentalists, for their part, see their tasks in almost Sisyphean terms. Brad Klafehn of the Colorado Open Space Council spoke not of dissension from within but of weariness from countering the persistent efforts of special interest groups who ''seek to weaken the new legislation and take the heart and soul out of the existing environmental regulations.''

Colorado may have the most extensive network of government agencies to ensure that coal development proceeds at a controlled pace and that benefits of that development accrue to the state. But its organization may also be the least rational and inconsistent. The state has opted for a strategy different from Montana's where coal development is not so much monitored as taxed, and taxed heavily. It is certainly too early to determine whether a single steep severance tax will restrain and regulate coal development more effectively than a myriad of overlapping regulatory agencies. Still, CEP suspects that if Colorado is sincere in its determination to control coal development and provide adequate benefits to the state, it may have to resort to a stiffer tax structure to succeed.

5. MONTANA

INTRODUCTION

The Fort Union Coal Formation lies in the Northern Great Plains, where ranching, small grain agriculture, and farming along the river valley are the primary elements of the regional economy. Industrial plants exist mainly in supportive roles—the processing of agricultural products, for example—and in and of themselves are not fundamental to the area. Towns serve as trade and transportation centers. Some coal has been mined in the past, but national attention to coal development here is recent.

National interest in Fort Union coal may be recent but it is now, for good reason, passionate. Stretching from northcentral Wyoming across southeastern. Montana and into North Dakota, the entire formation contains 1.5 trillion tons of coal—about 20% of the world's total coal reserves and 40% of the identified reserves in this country. Moreover, the coal commonly occurs in seams 20 to 30 feet thick, and occasionally in seams as thick as 250 feet. Much of it is so close to the surface that it is economically and technologically recoverable today by strip mining. As with other western coals, the sulfur content is low, between 0.2% and 0.7%, making the Fort Union area coal attractive for burning in coal-fired installations in the heavily polluted urban centers of the Midwest.

About half of this Fort Union coal lies in Montana. Nearly all is sub-bituminous in grade—not as high quality as the bituminous coal of Utah and Colorado but much better than the lignite reserves found in North Dakota and comparable to the coal of neighboring Wyoming. Montana's coal fields contain about 64% of all the sub-bituminous reserves in the West. Thirty-five billion tons of this coal can be strip mined at only a fraction of the cost of underground mining. Such extensive and accessible strippable reserves have aroused the interest of coal developers and turned them from the deep mined reserves of Utah and Colorado. Moreover, even in distant midwestern markets, Montana strip coal is now competitive with underground mined Appalachian coal.

Thus, Montana too can expect to be the focus of elaborate coal development schemes that could promise prosperity or threaten disruption of the traditional ways of life there. The coal industry already has expressed its interest in Montana coal by leasing over 600,000 acres of land from federal, state, Indian and private coal owners. Since the mid-1960s, several huge strip mines have begun cutting into the semi-arid sagebrush-dotted terrain in the eastern part of the state. Three of these mines, the Decker, Rosebud and Absalonka strip mines, are now the second, third and eighth largest coal mines in the country even though they opened only a few years ago. A direct Burlington Northern rail link from Montana to the Midwest has helped persuade several other companies to try to mine Montana coal for export east. Such energy conglomerates as Shell Oil and AMAX hope to have mines operating by the end of this decade. The presence of Pacific Power & Light among perspective coal developers suggests that Montana coal might soon be traveling West as well. Such extensive exports have fueled expectations that Montana coal production will reach 37.5 million tons by 1980, according to the Department of Interior. It is already reaching toward this goal: production soared from 3.5 million tons in 1970 to 27.3 million tons in 1977.

Forecasts such as Interior's suggest radical departures from the historically evolved patterns of Montana coal mining. All mining plans currently on the drawing boards call for strip operations in the southeastern part of the state, while most of the mining in the past has been underground in coal fields in northcentral Montana—most notably Bull Mountain, Great Falls and Lewistown Fields. Because these fields contained limited reserves of high sulfur coal in inaccessible, steeply dipping seams, mining in the past has been slight.

Minimal mining obviated any need for Montana to evolve a governmental infrastructure to systematically, comprehensively and reliably contain the adverse impact of coal development. Recently, however, Montana has styled itself as an advocate of slow, controlled, industrial growth. It has enacted the highest coal taxes of any western state including a severance tax that averages about 22% of the value of the mined coal. It has declared a moratorium on the leasing of state-owned coal land to complement the federal coal leasing moratorium. It has created extensive regulations affecting coal developers, established a myriad of agencies to implement these regulations, and granted agency staff sufficient power to enforce government policies.

Still, these precautions are new, hastily developed, and, some environmentalists suspect, deceptive in suggesting that the revenues accruing to the state and the strict regulations confronting developers will prevent environmental degradation. Whether or not Montana's manifest commitment to control the pace of coal development and meet the needs of coal-impacted areas is too little too late, however, remains to be seen. Its commitment is certainly impressive when placed against the lack of concern expressed by other western states.

FEDERAL LEASING

As in other western states, federal coal lands and coal leasing are key variables in Montana's future. Roughly 30% of the state's land mass covering 75% of the state's coal is federally owned. Only in Utah does the federal government control a greater percentage of coal. Montana's 22.6 million acres of federal coal lands are more than double the federal coal-bearing acreage of any other western state. Fully 35% of all federal coal lands in the West lie within Montana's borders.

Compared to most other western states, federal leasing activity in Montana has been limited. Only 17 coal leases spanning 36,232 acres of land, a scant 1.4% of federal coal land in the state, have been issued, leaving North Dakota as the only western state with less acreage under lease. These leases contain just over one billion tons of coal reserves, all surface mineable—only one-eighth of Wyoming's and one-third of Utah's federal coal reserves under lease.

There is more coal production on Montana's federal leases than anywhere else in the West, however. Six of the 17 leases are currently producing coal, giving Montana the highest percentage of active leases among the western states. Two of the mines operating on federal land, the Decker and Rosebud mines, are among the five largest strip mines in the country.

Perhaps because of this activity, corporate interest in federal coal leasing in Montana remains strong. Eighteen preference right lease applications are currently pending, spanning 26,000 acres and containing another 92 million tons of coal reserves. Since the federal leasing moratorium in 1973, more industry nominations for new lease acreage have concerned Montana land than land in any other state. Current leaseholders, for their part, seem content to hold on to their assets. Only two leases have changed hands by assignment since 1973, constituting a 12% lease assignment rate, much less than the western average of 25% during the same time period.

Resumption of federal coal leasing, then, is almost certain to encourage coal development rather than speculation in Montana. However, conflicts between surface owners of coal land and lessees of subsurface coal deposits continue to inhibit development of current leases and cloud in uncertainty the future of new leases. Ninety-seven percent of the land currently under lease in Montana is privately owned at the surface, often by successful ranchers whose families homesteaded the land decades ago. Only North Dakota has a higher percentage of surface–subsurface domain conflicts. Government policy on the leasing of federal subsurface mineral rights on lands whose surface is privately owned is uncertain.

INDIAN LEASING

Northern Plains Indians are the largest single ethnic group in the Northern Great Plains region. Among the various tribes that populate the area are the Crow and Northern Cheyenne, under whose reservations lie buried about 5% of the nation's coal reserves—a resource unknown to Congress when it forced the tribes onto the seemingly barren land about a century ago.

These two tribes are currently among the poorest in the nation, each experiencing seasonal unemployment of up to 70% in the winter. But coal development on either reservation or on adjoining land where the tribes retain mineral rights could bring an annual income of over $50,000 for tribal families now earning annually an average of less than $3,500. Each tribe, in its separate fashion, is currently preparing to use its vast and valuable coal resources to establish a power base and regain some of its lost prominence.

The adjoining Crow and Northern Cheyenne reservations span more than three million acres. Virtually all of the Northern Cheyenne reservation lies within the Fort Union formation and bears an estimated five billion tons of strippable coal. Only the easternmost five miles of the much larger Crow reservation contains strippable coal deposits, but reserve estimates here range as high as 10 billion tons. These 15 billion tons of coal buried under Indian land make development of Indian coal pivotal to Montana's coal future.

To date, the Bureau of Indian Affairs (BIA) has issued three Indian coal leases on the Crow reservation and one on Northern Cheyenne land. AMAX, Shell, and Westmoreland Resources control the Crow leases that span 75,000 acres. A 16,000-acre lease on the Northern Cheyenne reservation has been issued to Peabody Coal. An additional 16 prospecting permits have been issued, five on Crow and 11 on Northern Cheyenne land. Prospecting permit holders include Gulf Oil, Chevron Oil, and Consolidation Coal Co., a subsidiary of Continental Oil. Only the Westmoreland lease is now producing coal: its Absalonka mine is the eighth largest coal mine in the country. Westmoreland plans to expand production to 15 million tons per year by 1982.

The Westmoreland lease is also the only technically valid Indian lease or prospecting permit in Montana. The other contracts were invalidated in 1974 by then Secretary of Interior Rogers Morton. The Northern Cheyenne had requested the invalidation citing violations of leasing procedures by the Department of Interior and coal developers in the late 1960s and early 1970s. Morton agreed that the violations were so severe that all Indian tribes had the legal right to renegotiate or cancel existing contracts.

After three years of extensive negotiations, Westmoreland, alone among the various lessees, agreed to a new contract with the tribe. Signed in 1974, the new lease contains a royalty of 40¢ per ton of coal mined, a significant jump from 17.5¢ per ton authorized by the initial contract. The new lease also has an escalator clause calling for an increase in royalty payments in the event of a rise in coal prices. These provisions made the Crow the "highest paid coal owners in the country," according to Westmoreland president Pemberton Hutchinson.

The other Crow leases are bogged down in lawsuits, counter-suits and negotiations. Already the Crow have rejected a series of proposals from Shell and AMAX for providing inadequate compensation to the tribe. Divisiveness

112

within the tribe itself has also contributed to the stalled negotiations. While generally favoring development of its coal resources, the Crow have had difficulty agreeing on representative negotiators. Recently impeached tribal Chairman Patrick Stands Over Bull has formed the Crow Coal Council to rival the existing Coal Authority. The Authority favors renegotiating AMAX's lease with royalty payments on a percentage basis, while the Council prefers development on Shell's lease under a fixed royalty contract.

The Department of Interior has repeatedly attempted to break the deadlock in negotiations by threatening to cancel all leases outright unless substantial progress towards an agreement is made. Shell, for its part, claims to be negotiating with the Crow in good faith and has challenged Interior's intrusions into the court negotiations. Interior has repeatedly backed off from enforcing its ultimatums, leaving the courts as arbiter between the deadlocked Crow and the energy companies.

While the Crow have been fighting—often among themselves—over how to manage coal development, the Northern Cheyenne have been struggling towards a decision to forbid coal development on their land. The Northern Cheyenne seem more eager to protect their traditional livelihoods—agriculture and timber harvesting—and values from the disruption it fears from coal development. In 1975, they established the Northern Cheyenne Research Project to help the tribe "make their own mistakes and decisions," in the words of tribal Chairman Alan Rowland, "rather than allowing the federal agencies to make them for us."

In the Spring of 1976, the Northern Cheyenne requested the U.S. Environmental Protection Agency to designate their reservation a Class I area of pristine air in which no significant deterioration of air quality would be permitted. This request was motivated by Montana Power Co.'s plans to expand the nearby Colstrip power plant. The Northern Cheyenne claimed the pollutants emitted by the proposed Colstrip units 3 and 4 would aggravate the already high rate of respiratory illness and endanger wild plants they need for food, medicine and ceremonial uses. They are further concerned that construction and operation of the new plants would bring large numbers of non-Indians onto their reservation, reduce their numerical importance and dilute their language and culture.

A Class I designation would not only prevent the expansion of the Colstrip power plant but would also threaten future coal development projects in this portion of Montana. Fugitive dust emissions from strip mining activities on Northern Cheyenne land or power plant emissions from as far away as northeastern Wyoming could violate Class I standards. The designation is currently being challenged in the U.S. Court of Appeals. Until the issue is resolved—which could take years—there is an element of uncertainty about development projects here not found in most areas of the West.

STATE LEASING

More than in any other state, the history of state-owned coal leasing in Montana closely parallels that of the federal coal leasing program. Under both programs, most leases were issued non-competitively or at giveaway prices during the 1960s. Consequently, nearly all of the leased acreage is held by a small group of large corporations for speculative purposes. When a series of ad hoc attempts to remedy the earlier leasing mismanagement failed to quiet the local citizens, environmentalists and other government agencies calling for reform of leasing procedures, Montana declared a moratorium on state coal leasing in 1971—just as the federal government instituted its own moratorium. As the federal government has struggled to revamp its leasing program, Montana too has passed new leasing legislation and promulgated new leasing regulations. But Montana's efforts should pay off sooner and more sizably:while the federal government's proposed new leasing program has continued to draw fire from western landowners and environmentalists, Montana's new program is ready to go.

The Enabling Act of 1889 granted two noncontiguous sections in every township to the state government upon Montana's admission to the Union. Having sold nearly half a million acres since then, Montana's state government now controls 5.1 million acres. Most of this land lies scattered in the original single-section, 640-acre configurations, but because some of the sections had been homesteaded or included in adjacent Indian reservations, the state has been compensated with a few larger tracts.

The Montana Administrative Procedures Act of 1947 created the Board of Land Commissioners within the Montana Department of State Land to administer all activities on state-owned land, including of course, leased coal lands. The Land Board consists of five elected officials, including the Governor, State Auditor and Attorney General. In addition, a Land Commissioner appointed by the governor presides over the Board.

The oldest currently valid state coal lease was issued in 1963, 16 years after passage of the 1947 Act. Since then, the Land Board has issued 96 leases covering 51,947 acres of land (see Table 5-1). John Osborne, Administrator of Centralized Services in the Department of State Lands, divides the state leasing program's history into four distinct periods.

The first stage lasted from 1963 until 1965, when leases were issued according to the "open land" procedure. Any person or company licensed to conduct business within the state could request a lease tract. The Land Board would issue the lease for a $10 filing fee if the applicant could demonstrate that the acreage was not already under lease. Nine leases issued during this period.

In 1966 the state started to sell leases at public auctions via oral bidding. Lease tracts to be auctioned were selected from a list of nominations submitted by coal industry representatives. The first—and last—lease auction in Montana's history was held on May 7, 1966. According to Osborne, "it didn't turn

TABLE 5-1

MONTANA STATE LEASE ISSUANCE

Years	Number of Leases Issued
1963	1
1964	0
1965	7
1966	9
1967	22
1968	16
1969	0
1970	40
1971	1

out to well." It was, in fact, a disaster. CEP's research through the Montana archives revealed that 311 tracts of land were offered for lease, but bids were received for only 21. Of these bidders, 11 immediately backed down upon realizing that theirs was the only bid offered. Only eight leases were eventually issued. The average winning bid for this land was less than 50¢ per acre. Discouraged by the auction results, the Land Board reverted to the open land method and issued another 22 leases in 1967.

In mid-1967 the state revamped its leasing program to increase the revenues accruing to the state from coal leasing. Rental and royalty rates were raised. At the same time, the state decided to experiment again with competitive bidding. This time, prospective lessees would submit sealed bids opened at public auctions. No oral bidding would be permitted. This competitive bidding system was finally implemented in 1968. Fifty-six leases were thus issued at two lease sales—one in December 1968 and the other in June 1970. The sealed bid method proved more effective than oral auction, but was still less than spectacular in raising state revenues. Of the 56 tracts, only one bidder made an offer on each of 36 leases covering 22,795 acres. These bidders paid the state an average of $4.30 per acre. Two bidders appeared at 11 lease sales, offering an average of $13.10 an acre for each of the 5,661 acres leased. The remaining 3,797 acres were sold in nine leases, each of which interested three bidders. The winning bidders on these leases paid an average of $13.80 per acre to outbid competitors.

Since issuing the first lease in 1963, the state has received $285,696 or $2,976 for each of the 96 coal leases issued (see Table 5-2). This is the second highest average among the six states examined.

Competitive lease sales, however unspectacular, have accounted for more than 70% of all state revenues associated with the issuance of coal leases. The 64 leases issued competitively—eight via oral and 56 via sealed bidding—have

TABLE 5-2

REVENUE FROM MONTANA STATE LEASE SALES

Method of Leasing	Number of Leases	Total Revenue Collected
Open Land	31	$310
Segregation	0	0
Lease Exchange	—*	77,118
Special Preference Right	1	1,064
Competitive Bidding	64	207,204
TOTAL	96	$285,696

*19 leases issued prior to 1967 were exchanged for new terms in 1971; no new leases were created.

sold for an average of $6.32 per acre, the second highest average among the western states conducting competitive lease sales. Cash bonus payments also have been second highest in the West, totalling $207,204.

A second source of lease issuance revenue has been the filing fees collected by the state to defray the costs of issuing open land coal leases. Thirty-one open land leases have produced $310.

In 1971 Montana required that holders of the 19 leases issued prior to the mid-1967 leasing revisions exchange their lease contracts for newer ones with higher royalty rates. The state collected special "exchange fees" of either $5 or $10 per acre, depending on the lease. On September 15,1971 the old leases were reissued with new terms, and the state collected $77,118 in exchange fees.

Montana's granting of a special preference right lease to Decker Coal Co. in August 1971 provided the state's final source of lease issuance revenue. Decker claimed to need the small 8.5-acre tract adjoining its existing state and federal leases to continue strip mining operations at its mammoth Decker mine. Decker offered to pay the state a cash bonus of $125 per acre to avoid having to divert the mine around that tract. Since obtaining the lease, Decker has mined through the land and extracted roughly 300,000 tons of coal before returning to adjoining land covered by other leases. This was the last lease issued before the moratorium.

All valid coal leases in Montana last for 20 years during which time lessees are required to pay rent to maintain their leases. Different lease contracts contain different rent provisions. Total rent collections in Fiscal 1977 were $87,488, or $1.68 per acre per year, slightly higher than the average rental rate in other western states.

Only one state lease in Montana is now producing coal. Based on the 5.1 million tons produced on this 640-acre tract last year, Decker Coal Co. paid more than $865,000 in royalties. This payment reflects a royalty rate of 17.5¢ per ton. Even though this was the only producing state lease in Montana last year, Decker's mine achieved a level of production that far exceeded total output from state-owned land in every other western state. Only one other state coal lease in Montana, the special case preference right lease also controlled by Decker, has ever produced a single ton of coal.

The 96 valid state coal leases in Montana are distributed among 16 companies. This average of six leases per lessee represents the highest level of lease concentration in the West. As in the other western states, though, here too the top five leaseholders control most of the leased land. The five largest leaseholders account for 68% of the leases and 65% of the acreage under lease. The top two leaseholders, both subsidiaries of major oil companies, control nearly 50% of the leased land in the state. Consolidation Coal, a Continental Oil subsidiary, owns 29 leases covering 15,479 acres—a whopping 30% of all leased acreage. Exxon subsidiary Carter Oil Co. ranks second, with 17 leases spanning more than 9,000 acres. Other top leaseholders include Peabody Coal Co., Peter Kiewit Sons, and Decker Coal Co., which Peter Kiewit Sons owns in a joint venture with Pacific Power & Light (see Table 5-3).

Low lease maintenance revenues, low coal output and high lease concentration have continued unabated even through Montana's moratorium on state leasing. That these existing leases represent a stumbling block to coordinated energy development within the state is further underscored by the state's failure to prevent or regulate the bartering of leases on the open market. Montana law permits leaseholders to sell their leases to anyone they please and to keep any cash payments or overriding royalties included in the transaction. CEP found that, while no new leases have been issued since the moratorium was instituted in 1971, 35 leases representing 36% of the state's total have changed hands in the last six years. Only one other western state has experienced a higher rate of assignment. In a startling example of the kind of lease activity that Montana's moratorium has permitted, Carter Oil Co., the state's second biggest leaseholder, obtained all 17 of its leases by assignment on one day in July 1975.

TABLE 5-3

TOP FIVE MONTANA STATE LEASEHOLDERS

Leaseholder	Number of Leases	Acreage under Lease	% Total Acreage under Lease
Consolidation Coal Co.	29	15,479	29.8
Carter Oil Co.	17	9,495	18.3
Peabody Coal Co.	6	3,837	7.4
Peter Kiewit Sons Co.	5	2,639	5.1
Decker Coal Co.	8	2,568	4.9
TOTAL	65	34,018	65.5

The legacy of the past notwithstanding, the Montana Department of State Lands and the Board of Land Commissioners are now anxious to implement a new leasing program, the product of a law passed by the state legislature in 1975. CEP believes that the new leasing program, if properly enforced, is exemplary among the six western states examined.

The new program will cause dramatic changes in lease issuance procedures. Open land leasing will be abolished in favor of competitive leasing, which the Department of State Lands will regulate much more actively than it has in the past. Industry nominations of lease sites will still be accepted, but no tract will be leased until the state has evaluated its resources, estimated the value of the coal reserves covered by the lease, and assessed the environmental impact of mining the tract. The Department will also propose tracts for leasing on its own. Here too the guidelines for lease proposals are specific, directing the state not to lease lands where strip mining would prevent the eventual extraction of underground reserves, to avoid leasing land whose present surface usage would be threatened by mining, and to lease only compact, mineable units of land.

The new legislation gives the State Land Board extensive discretion in the management of future lease sales. The Board may issue the lease to the bidder offering the highest one time cash bonus, or it may choose to employ variable royalty bidding, whereby the lease would be awarded to the applicant offering to pay the highest royalty rate to the state. The Board may choose either oral or sealed bidding, or it can reject any and all bids if they fall below the "fair market value" of the land as determined by the resource evaluation.

Leases will last for 10 rather than 20 years according to the provisions of the new program. Leases may not be renewed unless coal is being produced from an active mine at the lease expiration date, or the lessee has obtained a mining permit or mine site location permit from the state. Such permits will not be issued unless the lessee submits concrete plans to open a mine. These measures should stem the speculative holding of state-owned land.

Royalty rates for new leases will vary—especially if variable royalty bidding is employed as the lease issuance procedure—but they may not fall below 10% of the gross value of the coal. Rent is established at $2 per acre, nearly twice the current level. Rental payments may not be credited against royalties, as they are under most state leasing programs in the West. They must, therefore, be paid whether coal is produced or not. Advanced royalties, to be credited against possible actual royalties owed only during the year of the advanced royalty payment, are required of all non-producing leases. Most western states allow advanced royalties to be charged against future coal production occurring any time over the lifetime of the lease. Royalty, rental and advanced royalty rates may be adjusted at five-year intervals after the initial 10-year lease term.

Future lessees may still assign their leases, but the state is empowered to reject any assignment when "the interests of the state" would be "prejudiced." A state refusal can be appealed in court, however.

Revisions in leasing regulations promise to close some of the gaping loopholes of the earlier state leasing program, but without frustrating the aspirations of prospective lessees. Ethel Schenck of the Department of State Lands told CEP that new legislation notwithstanding, 60 lease applications were pending before the Land Board at the end of 1977. The processing of these applications will require the Board to weigh the conflicting directives of the new lease regulations—which calls for increased resource planning, and the Enabling Act of 1889—which instructs the state to manage its land to benefit 12 Congressionally-selected institutions within the state (see Table 5-4).

All revenue derived from activity on a particular institution's land must become a part of that institution's operating budget or trust fund. Rent payments become part of the operating budget of the institution controlling the acreage for which the rent was paid. Royalties and advanced royalties are deposited in the State Permanent Trust Fund and invested; each institution is awarded only its share of the interest from these investments. Montana's Permanent Trust Fund now has a balance of $79 million, making it the second smallest fund in the six states surveyed.

Total income from all activity on state land has risen dramatically, from $8.3 million in 1965 to $22.6 million in 1976. Rent and royalties from coal

TABLE 5-4

BENEFICIARIES OF MONTANA STATE LANDS

Institution	Acreage	Fund Balance
Public School	4,595,800.81	$71,653,649.30
University of Montana	18,160.87	943,713.11
Montana State University*	95,385.62	2,292,710.03
Montana College of Mineral Science and Technology	59,606.22	1,513,638.69
State Normal School	62,890.00	1,294,005.08
Deaf and Blind Asylum	36,235.86	628,967.81
State Reform School	68,744.01	650,015.98
Public Buildings	186,227.08	—
Veterans Home	1,275.61	8,398.47
Militia Camp**	640.00	—
Agricultural and Manual Training School	5,000.00	—
State Penitentiary	9.75	—
TOTAL	5,129,973.00	$78,976,696.47

*includes both Morrill Grant and Second Grant
**now used as an Agricultural Experiment Station

development contributed roughly $1 million or 5% of the 1976 total—a much higher percentage than in any other western state.

As activity on state coal land has generated increasing revenues, the Montana Land Board has attempted to redefine its function, from pure revenue maximization and investment counseling to the institutions, towards resource planning for the state. Alone among the western states, Montana's Board has carved out a policy to guide it in cases where the issuance of coal leases and approval of coal mining would be beneficial to the institutions but, because of environmental degradation, would run counter to the interest of the state. The new lease regulations authorize that "in the exercise of its power, the guiding rule and principle of the Board of Land Commissioners is that the state lands are held in trust for the support of education and for the attainment of other worthy objects helpful to the well-being of the people of the state; and that it is the duty of the state to so administer this trust as to secure the largest measure of legitimate and reasonable advantage to the state." By equating the interests of the 12 institutions with that of the state, the Land Board hopes to fulfill the mandates of the Enabling Act and the new leasing legislation. Both as an investment counselor and resource manager, the Board of Land Commissioners will ultimately be serving the interest of the people of Montana.

DEVELOPMENT

Montana is unquestionably experiencing a coal mining boom. Statewide production has tripled since 1973 to more than 27 million tons in 1977. The Northern Great Plains Resource Program's intermediate forecast for coal production expansion projects Montana coal output to triple again by 2000, a rate of expansion even faster than that forecast for booming Wyoming.

A handful of massive strip mining operations dominate Montana's coal production, each having either opened or vastly expanded operations in the last four years. The Decker Mine, operated by Pacific Power & Light and Peter Kiewit Sons in a joint venture, is currently the second largest coal mine in the country, producing more than 10 million tons a year. Montana Power Co.'s subsidiary, Western Energy, operates a strip mine in Rosebud County only slightly smaller. 1977 output here was 9.7 million tons, making this the third largest coal mine in the country. Westmoreland Resources, a consortium of Westmoreland Coal, Morrison-Knudson, Kewanee Oil and Penn-Virginia, operates the eighth largest coal mine in the country on land covering reserves leased from the Crow tribe. Production at Peabody Coal Co.'s Big Sky mine continues to hover around three million tons per year.

The companies operating these mines hope to further expand production in the next decade. The Decker mine alone could be producing 35 million tons of coal annually by 1985—more than the entire state's output in 1977. Meanwhile, several other corporations are scheduled to open new mining operations. Shell, for example, plans four mines on Crow land covering 1.3 billion tons of recoverable reserves. Cumulative production from these operations could total 30 million tons of coal per year by 1990. AMAX too plans a large mine on Crow land, east of the Sarpy Creek. If Colstrip power plant units 3 and 4 proposed by Montana Power Co. pass, Environmental Protection Agency, (EPA) review, Western Energy Co.'s Rosebud mine may double production to 19 million tons annually in the 1980s. Pacific Power & Light's Spring Creek mine in Bighorn County and Burlington Northern's Circle West mine in Mc-Cone County complete the list of expansive strip mining operations slated for southwestern Montana.

Many of the nation's energy conglomerates have flocked to Montana this decade, not only because of the accessibility of strippable reserves, but also because of the ease of transporting them to major markets in the Midwest. The Burlington Northern Railroad ships Montana coal directly to such Great Lakes area utilities as Detroit Edison, Commonwealth Edison, Dairyland Power Cooperative, Wisconsin Power & Light, and Northern States Power Co. Also hoping to utilize this transportation network, American Electric Power—a holding company comprised of several power companies in Indiana, Kentucky, Michigan and Ohio—plans a captive mine in Montana to supply coal to power plants as far away as the Ohio River Valley.

Montana's coal could also be exported to far western markets in the near future. Washington Power Co. and Washington Gas Co.'s subsidiary, Thermal Energy Inc., have recently joined together with ITT to develop Montana coal reserves in a manner similar to American Electric Power. Portland-based Pacific Power & Light is among Montana's most prominent leaseholders. The use of Montana coal for power generation in the Pacific Northwest could be further boosted by Congressional passage of Senator Henry Jackson's (Dem.-WA) Pacific Northwest Electric Power Supply and Conservation Act. This bill proposes to expand the authority of the Bonneville Power Administration (BPA), a quasi-public entity similiar to the Tennessee Valley Authority, to build, operate and finance the construction of coal burning power plants.

BPA was formed in the 1930s to develop the enormous hydro-electric potential of the Columbia River and its tributaries. It has provided cheap electricity to industries in northwestern municipalities for nearly 40 years. The hydro-electric potential of the water system is almost completely tapped now, however, and the BPA is anxious to adapt new electrical generating technologies to meet the increasing demand for electricity within the region.

Jackson's bill would allow BPA to build or assist other companies in building coal burning power plants. The Authority would then be permitted to combine the cost of hydro-electricity, which is still very low, with the projected higher cost of generating electricity by burning coal, and offer its customers an average rate still generally cheaper than electricity from other sources. Furthermore, BPA's use of its tax-exempt bonding capacity to build, or help finance other companies to build, coal-fired power plants could provide another strong incentive for rapid coal development expansion in Montana.

STATE POSTURE

The large strip mining companies and the midwestern and Pacific northwestern utilities that are committed to the development of Montana's coal reserves suggest an uncontrolled boom on the horizon. State agencies and officials, however, have joined with the ranchers and river valley farmers in southeastern Montana to maximize the limitations and safeguards exerted on coal development in the state. More than any other western state, Montana is officially and actively committed to slow, controlled coal growth.

Various impulses have helped shape this commitment. Montana's lands are among the most fertile in the West, supporting a prosperous agricultural economy whose importance can only be diminished by rapid coal development. The state, therefore, can better afford to move more slowly than neighboring Wyoming. Montana is also the seat of a vocal environmental movement that scrutinizes public and corporate policy and urges comprehensive restraints on development.

Underlying the concerns of both ranchers and environmentalists is a shared distrust of the major mining companies, which have traditionally been a significant political force in the state. Many feel that these companies have not always worked in the public interest. As recently as 1961, the Montana legislature, under heavy industry pressure, declared mining a "public use," giving mineral companies the right of eminent domain on lands with surface-subsurface ownership conflicts. "Episodes of angry confrontation and near violence multiplied," according to Alvin Josephy, a writer for *Audubon* magazine, "as the purchasers—nervously eyeing the progress of competitors and aware of large secret corporate plans that depend on timely acquisitions—pressured the landowners." The recent history of such incidents has helped provoke Montana's bitterness towards rapid uncontrolled coal development.

A cautious posture towards development is reflected in Montana's coal tax structure—the most elaborate and effective tax system in the West. Layering three different taxes with different rates, Montana now assesses higher coal-related taxes than any other western state. Coal companies are paying nearly $1.50 in taxes for every ton of coal mined in 1977. Annual revenues from coal taxes now exceed $40 million.

The Resource Indemnity Trust Tax, instituted by voter initiative in 1971, was the first tax passed in Montana to specifically affect coal developers, though it also applies to uranium, oil and natural gas extractions. Its rate is low—$25 plus 0.5% of the gross value of production in excess of $5,000—because, according to Alan Davis, a research analyst for Montana's Department of Revenue, it was established "before full scale development broke out."

Revenues generated by this tax are placed in a trust fund ultimately to be spent for environmental protection and improvement. No money can be appropriated until the fund's balance reaches $10 million. Then, only the interest income may be spent. When the fund reaches $10 million, income above that as well as interest may be expended, but the $10 million remains a permanent endowment to future generations of Montanans. Since the fund's balance is still less than $8 million, no money generated by this tax has been spent to date.

As in other western states, a property tax is also assessed Montana coal developers at the county level. The rate of property taxation for coal developers underwent a significant revision in 1975. The old property tax was based on the net proceeds of each mine's output. The new assessment is based on 45% of the gross proceeds of strip mining operations and 33.3% of the gross proceeds from underground mines. Since the new property tax rates have taken effect, taxable value of Montana coal has jumped by 125%, from $23 million in 1975 to $53.2 million in 1977. Davis, however, insists that even though the revision has "made the counties pretty well off," the steep increase in taxable value is more directly attributable to the 94% jump in state coal production and the steady increase in the price of coal over the same time period. The change from net to gross proceeds was more an administrative reform to make the levying of the tax less complicated than an effort to further discourage coal developers. Still Davis does allow that the new tax will insure a more stable and reliable flow of revenue to the counties, if not necessarily a more sizeable one.

The coal severance tax which replaced the old strip mine coal license tax, has proven to be Montana's most effective coal-related source of revenue. In Fiscal 1977, the state collected nearly $36 million through its severance tax, many times more than the severance tax collections of any other western state.

The original license tax on strip coal, enacted in 1921, established the taxation rate at 5¢ on every ton in excess of 50,000 tons extracted from each mine. This rate crept upward until 1973, when the state was charging as much as 40¢ per ton for the privilege of extracting strip coal The state legislature repealed the license tax and instituted a severance tax in 1975. Severance tax rates vary according to the quality of the coal mined and the mining method employed. Coal companies are now paying an average of 22.5% of the contract selling price of the coal they mine in Montana.

The severance tax has provided the state with steadily increasing revenues. More money was collected in its first year of operation than was generated by the old license tax in the previous five years combined. The Department of Revenue expects these increases to continue, projecting $51 million in taxes in Fiscal 1979—40% more than the previous year's yield.

The allocation of severance tax revenues is subject to constant revision. Originally, the state's general fund received 40% of severance tax collections. Its share will be scaled down gradually until 1980, when it will be apportioned 19.5% of tax revenues. Allocations to the school equalization fund, the county land planning agencies, and the local impact and educational trust, which exceeded 28% of tax revenues in 1977, will also be gradually reduced. Through 1979, 9.75% of severance tax revenues will go for highway construction in areas heavily affected by coal development. This allocation will then be eliminated.

Apparently, the major beneficiary of the new severance tax allocation formulas will be the recently established coal tax trust fund. The trust fund was authorized by a voter initiative in 1976 amending the original severance tax allocation legislation. The fund will indemnify the state against damages from coal development even after the coal is gone—thereby complementing the "pay as you go" spirit that underlies the severance tax. The new allocation formulas apportion 50% of all severance tax revenues to the coal tax trust fund in 1980. Its balance will eventually stretch into the hundreds of millions of dollars. Because this fund will keep so much money from state officials and politicians eager to spend it on special projects, it is not a very popular addition to the coal tax picture—at least within the state government. As Alan Davis complained to CEP, the trust fund will "lock up lots and lots of our money forever and ever."

But while the severance tax has not discouraged development as was originally forecast, it has provoked resentment among mining companies and their customers. On June 20, 1978, 11 midwestern and southwestern utilities led by Commonwealth Edison, Detroit Edison, the Lower Colorado River Valley Authority, and the city of Austin, Texas, joined their suppliers, Decker Coal Co., Peabody Coal Co., and Westmoreland Resources, in challenging the severance tax in Montana's state courts. Claiming that the tax rate bears no legitimate relationship to the service or protection provided by the state, these utilities and coal companies seek to recover their first quarter 1978 tax payment of $5.5 million—which they begrudgingly paid under protest.

6. NEW MEXICO

INTRODUCTION

Coal mining in New Mexico dates back to 1860 when coal fields in the central part of the territory supplied the U.S. Cavalry. Later, the construction of the Southwest's first railroad network linked New Mexico territory's coal fields to new markets and provoked more sustained coal mining. Annual state output increased from 25,000 tons in 1887 to over four million tons in 1917—a level of production not surpassed there until 1969.

Throughout much of the twentieth century, energy developers within the state overlooked coal in favor of other energy resources. Oil and gas exploration once dominated the energy industry's involvement in New Mexico. Later, developers discovered that 40% of the country's uranium reserves lie buried under New Mexico lands, and made uranium development the cornerstone of the energy industry in New Mexico.

Recently, New Mexico has begun to share in the western coal boom too. The growing demand for steam coal by electric utilities in Arizona, southern California, and within New Mexico itself helped revive the state's coal industry in the 1960s. The advent of large scale strip mining operations in the northwestern corner of the state increased output. For example, the opening of the state's first strip mine in 1962—Gulf subsidiary Pittsburgh & Midway Co.'s McKinley mine, supplying coal to Arizona Public Service Co.'s Cholla plant—increased statewide coal output by 195% over the previous year. Utah International opened its Navajo strip mine in 1963 and gradually expanded its output to a capacity of eight million tons per year. Until 1974, this was the largest coal strip mine in the world. Most coal mined here was burned in the adjacent Four Corners power plant.

In addition to the McKinley and Navajo strip mines, there are now four other active mines of significance in New Mexico. Each either opened or greatly expanded its production since 1960. Kaiser Steel Corp. operates two; Western Coal Co., a subsidiary of New Mexico Public Service Co., operates one; and AMCOAL operates one.

These six mines contributed nearly all of the 11,644,000 tons of coal produced in New Mexico in 1977. This level of production ranked New Mexico third among the western states in annual production, behind Wyoming and Montana.

The pattern of a few large mines dominating the state's coal industry, though consistent with other western states, is a new development in the history of the New Mexico coal industry. Most of the 221,250,000 tons of coal which have been mined in the state to date have come from an abundance of small, inconsequential mining operations. In 1955, for example, 31 different mines contributed to New Mexico's coal output of 202,000 tons. Average production per mine has shot up since then by a factor of 300, from 6,500 to almost two million tons per year.

Though revived, the New Mexico coal industry still contributes less than 2% of the nation's total coal production. The magnitude of future expansion remains unclear. Only a few large mining projects for federal land have been announced. The fate of several huge strip operations now on the drawing boards for the Navajo Indian Reservation depends upon the commercialization of coal gasification technology and the energy resource decisions of the tribe, which may ban gasification plants from the reservation and limit future coal projects to those owned and operated by Navajos.

New Mexico's coal reserves are impressive. Of the state's total land area, 25,000 square miles, or 20%, contain coal. Buried underneath these coal lands lie more than 2.5 billion tons of sub-bituminous coal and another two billion tons of higher grade bituminous reserves. Roughly 80% of the sub-bituminous reserves lie close enough to the surface to be strip mined. Sulfur content of these strippable reserves is often as low as 0.4%.

Most of the state's low sulfur, strippable, sub-bituminous coal lies in the San Juan River Basin in northwestern New Mexico. Spanning 26,000 square miles, the basin contains vast coal, uranium, and natural gas deposits. Its 19 different coal fields contain roughly 90% of the state's coal resources. Most current and planned production occurs in two of these fields, the Fruitland and Mesa Verde coal fields.

There are nine additional coal fields scattered across New Mexico. The most important is the Raton field in the northeastern portion of the state, where New Mexico's highest quality coal—14,000 BTUs per pound—is found. Only 15% of the reserves there are considered accessible by modern mining methods, however. Kaiser Steel operates two mines in this field.

Steeply-dipping seams, erratic deposits, high ash content and other geological problems hinder development of New Mexico's other coal fields.

FEDERAL LEASING

The federal government owns 59% or 5.5 million acres of the 9.4 million acres of coal-bearing land in New Mexico. Less than 1% of federal coal lands in the state are currently under lease. The 28 valid coal leases span 40,959 acres, second lowest leased federal acreage among the six western states. Leased reserves total 335 million tons, less than in any western state except North Dakota. Roughly 83% of New Mexico's leased federal reserves are strippable.

No new leases have been issued nor any relinquished since 1973. In 1977 only four leases produced any coal. Total production was 2,849,042 tons, the second lowest level of federal production among the western states. Since the leasing program was instituted in 1920, fewer tons of coal have been mined from federal land in New Mexico than anywhere else in the West.

Still, federal coal activity in New Mexico is on the upswing. Federal production, though still small compared to other western states, has increased fivefold in the past five years. Though no new leases have been issued, 17 of the 28 leases have changed hands by assignment since 1973. This 61% assignment rate is by far the highest in the West. There are 26 pending preference right lease applications spanning more than 75,000 acres and covering roughly 500 million tons of recoverable coal.

Although many coal developers clearly recognize the value of New Mexico's federal coal deposits, they seem to have limited aspirations. The Bureau of Land Management (BLM) estimates that federal coal production will increase to about 5.3 million tons in 1990—not even double the production in 1977. This gradual expansion contrasts starkly with the tenfold increases projected for federal coal output in Utah and North Dakota and the threefold increase forecast for federal lands in the West at large.

INDIAN LEASING

The Navajo Indian Reservation covers 16 million acres, roughly the size of West Virginia, and stretches across northwestern New Mexico, northeastern Arizona and southeastern Utah. The climate there is dry and the lands are mostly desert. Water is scarce and precious; average annual rainfall seldom exceeds six inches. The largest Indian tribe in the country, the Navajo nation has traditionally survived on minimal sheep and cattle raising and subsistence farming.

Coal could change all this. The 2.9 million acres of Navajo land within New Mexico's borders contain between two and four billion tons of recoverable coal reserves. Navajo coal is clearly the big ticket item in the New Mexico coal development industry.

The tribe has issued three huge coal leases covering parts of the richest coal-bearing land on the reservation. These three leases span more than 82,000 acres of Indian land, more than double the acreage contained in New Mexico's 28 valid federal coal leases. The average Navajo lease is 19 times larger than the average federal lease in New Mexico.

Utah International controls the oldest Navajo lease. In 1957 it obtained rights to a 31,416-acre tract, site of the mammoth Navajo strip mine. Production there accounted for almost two-thirds of statewide coal output in 1977. Consolidation Coal Co. and El Paso Natural Gas Co. control a second lease tract covering 40,281 acres and more than 800 million tons of recoverable reserves. The companies hope to gasify coal mined there in their controversial CONPASO plant, also planned for the reservation. The smallest of the three Navajo leases is a 11,157-acre tract issued to Pittsburgh and Midway in 1964, adjoining the federal lease where the McKinley mine had been developed. Roughly 87,000 tons of coal were mined from the Navajo portion of the mine last year.

Further development on the reservation has provoked much controversy. The National Indian Youth Council and 13 residents of the Navajo community of Burnham filed a lawsuit in February 1978 to halt the planned strip mine on Consolidation Coal Co.'s and El Paso Natural Gas Co.'s lease. The Navajo plaintiffs claimed that the proposed strip mine—an integral part of the CONPASO venture—would require the relocation of 200 Navajo livestock ranchers, destroy native plants and herbs used in healing ceremonies, threaten rich archeological resources, and disrupt the spiritual and physical traditions of Navajo life. The lawsuit is directed against the Department of the Interior's Bureaus of Indian Affairs and Reclamation and accuses them of failing to protect Indian land and resources.

Consolidation Coal and El Paso had hoped to feed coal from this strip mine into two of six even more controversial gasification plants. These would be the nation's first, and the world's largest, commercial gasification ventures. The other four controversial gasification plants are on the drawing board of the Western Gasification Co. (WESCO), a joint venture of Pacific Lighting Corp. of southern California and Texas Eastern Transmission of Houston, Texas. WESCO's plans have also run afoul of the Navajo. The company had hoped to fuel its plants with coal strip mined at Utah International's Navajo mine. Five years of corporate-tribal negotiations reached an apparent definitive end on February 1, 1978 when the Navajo Tribal Council rejected by a vote of 48 to eight WESCO's proposal to build its four gasification units—despite WESCO's promise to include the tribe as a limited partner in the venture.

Navajo Tribal Council Chairman Peter MacDonald has led the assault on WESCO and CONPASO, claiming that "coal gasification has not proved itself." He and other opponents of the plants charge that the Lurgi gasification technology, which each proposed plant would employ, is costly and inefficient. They also fear that the plants would cause high levels of air pollution, releasing toxic chemicals such as lead, mercury and boron; consume thousands of acres of Navajo grazing land; and divert critical quantities of water from the new Navajo Indian Reservation Irrigation Project. To further document their dissatisfaction, they cited the influx of non-Indians onto the reservation—45% of all construction jobs would be filled by non-Navajo personnel as would 95% of the permanent jobs—and the appearance of boom towns which they claimed would disrupt the traditions of tribal life. The compensation seemed inadequate. WESCO's original proposal, for example, called for the company to pay the tribe 0.5% of its gross revenues, or about $2 million a year. Even the Department of Interior suggested 3–5% as more equitable, but the company rejected their suggestion.

WESCO has developed an alternative plan to build a gasification plant just beyond the reservation's perimeter. From an off-reservation site, WESCO would pay the tribe $100,000 for permanent right-of-way privileges to transport the coal from the Navajo mine to the plant site, $50 an acre for the little land it would still need to lease, and a resource utilization fee for the right to use Navajo coal from the Utah International mine.

Despite the lawsuit facing CONPASO and the Tribal Council rejection stinging WESCO, both of these companies seem intent on pushing ahead with projects on which they have spent over $50 million and several years of planning. No matter what the outcome of tribal negotiations, however, large technical and economic problems still hamper the synthetic fuel industry. Gasification schemes have not been tested in this country on a commercial scale and government policies on the pricing of synthetic fuels are still uncertain. These gasification schemes could fall flat yet.

STATE LEASING

"Remember," cautioned Jack Kennedy, Director of the Minerals Division of the New Mexico Division of State Lands, "this office does not operate for the health and welfare of the people of New Mexico." His tone spoke not of failure but of success. Like state land offices in every other western state, Kennedy's office does not seek to plan coal development for the state, generate revenues to public coffers or limit environmental degradation. Instead, its task is to raise money for a limited number of "beneficiary institutions."

This land management perspective dates back to the Ferguson Act of 1898 when the federal government granted to the public schools two sections of land in every township in the newly-surveyed territory. Additional institutions and acreages were authorized by the Enabling Act of 1910, which preceded New Mexico's entry into the Union. The state once controlled more than 13 million acres. Some have been sold or exchanged, reducing current state ownership to 10 million acres. Meanwhile, the number of beneficiary institutions has swelled to 20, including the public schools, universities, mental hospitals, and the state military institute (see Table 6-4, page 134).

The Enabling Act invested in the Commissioner of Public Lands the responsibility to promote and supervise commercial activity on state-owned lands to raise money for the state institutions. Only the 20 institutions were to benefit from revenue-generating activity. Any diversion of these monies for other purposes was considered a breach of trust.

CEP found no other state land office which interpreted its responsibilities so rigidly. When a CEP researcher arrived in Santa Fe to review state leasing files and interview government officials, Director Kennedy refused to be interviewed. Carlos Gallegos, Executive Assistant Commissioner, added that "We don't do research for anybody." Established with a "very restrictive" mandate, the division was not authorized to use its funding to "help compile data for CEP." When at the conclusion of a short interview with Gallegos several questions remained unanswered, Gallegos risked a minor breach of trust himself by cajoling Kennedy into joining the interview. Kennedy walked in and quickly cautioned that he could not speak, "without first checking with the wards of his trust."

This meticulous attention to detail has infused the state leasing program in New Mexico with a morale, energy and sense of duty not found elsewhere in the West. Consequently, many aspects of New Mexico's program are "better" than in other states; that is, they generate more revenue and more competitive interest, and better facilitate coal development.

The failure in New Mexico, however, is that attention is lavished on only a few aspects of the leasing program, while large areas of legitimate responsibility are left unattended. For example, the Division of State Lands has never obeyed the Enabling Act directive that "all lands before being offered, shall be appraised at their true value...and no sale of, or other disposal thereof, shall be made for less than the value so ascertained." Instead, the state responds to industry nominations of tracts it seeks for leasing. The state does no drilling or exploration prior to offering a tract for lease because, in Kennedy's words, "it is too expensive." No effort is made to estimate a minimum bonus bid or determine competitive interest in the land. Kennedy assesses state land by "taking a look at the maps and if the land is open, putting it up for lease." Consequently, the state has no idea how much coal it owns or where it can be found. When asked to estimate the coal reserves on the 10 million acres of state land in New Mexico, Gallegos told CEP, "Whether there is any coal at all is strictly a guess."

The extent of state coal ownership is not a guessing game for everyone. The Southwest Research and Information Center, an Albuquerque-based public interest group, estimates that the state is the outright owner of 15.3 billion tons of coal reserves, an evaluation that elicits only a shrug from Kennedy.

The interest of coal developers in obtaining state coal leases in New Mexico suggests that they too know more about New Mexico's state coal reserves than the Division of State Lands' Mineral Division. The Division of State Lands has issued 218 currently valid coal leases spanning 106,860 acres. Each lease is less than 1,000 acres, reflecting the checkerboarding of non-contiguous state-owned tracts.

Leasing in New Mexico is a recent phenomenon. Only 12 leases were issued before 1970. In the past five years, fully 89% of all currently valid leases have been issued (see Table 6-1).

CEP's analysis of the New Mexico state leasing program revealed two distinct eras of lease issuance. Until 1972, coal leases were issued according to preference right procedures. Any applicant who could locate unleased land

TABLE 6-1

NEW MEXICO STATE LEASE ISSUANCE

Years	Number of Leases Issued
1960-1961	1
1962-1965	0
1966-1967	1
1968-1969	10
1970-1971	12
1972-1973	39
1974	27
1975	0
1976	39
1977	89

TABLE 6-2

REVENUES FROM NEW MEXICO STATE LEASE SALES

Method of Leasing	Number of Leases	Total Revenue Collected
Preference Right	24	$920
Segregation	0	0
Competitive Bidding	194	$724,572
TOTAL	218	$725,492

tracts on state land maps could receive a prospecting permit upon payment of a fee ranging from $10 to $25 depending on acreage. Prospecting permits could be converted into coal leases if the permittee discovered commercial quantities of coal within one year. Preference right procedures accounted for the issuance of 24 leases from 1960 to 1971, bringing the state $920 in total lease issuance revenues (see Table 6-2).

Competitive leasing was instituted in 1973. Since then, 194, or 89%, of all outstanding coal leases have been issued at five competitive sales. CEP found that the 97,783 acres leased competitively have brought an average bid of $7.41 per acre. In no western state have more acres been leased competitively at so high a price. Revenues from competitive lease sales total $724,572, again the highest in the West.

Although New Mexico has exceeded the standards set in other western states for revenues collected from competitive lease sales, the standards themselves are low. Existing evidence suggests that the lease sales to date have been less than rigorous in their pursuit of revenues. The Southwest Research and Information Center studied the first four competitive lease sales and found that 88% of the 2,050 tracts offered for lease drew no bidders. Of the remaining 284 tracts, 215 attracted only one bidder. Genuine competition was limited to the 50 tracts which attracted two bidders and the 19 which attracted three or more.

When CEP visited the offices of the Division of State Land in October 1977 to confirm the findings of the Southwest Research and Information Center, we were told that records of the first four lease sales had been discarded. Since another lease sale was scheduled in November 1977, however, CEP was afforded an opportunity to observe New Mexico's competitive leasing system in operation. The results were conclusive. Two hundred and ninety-nine tracts of land were offered for lease. No bidders appeared willing to put up cash for 209 of them. Only one bidder appeared to offer a cash payment for 71 of the 90 leases which were eventually issued. Any semblance of competitive bargaining for leases was restricted to the remaining 19 leases for which two bidders bartered.

CEP believes this sale, like the others, represents a near giveaway of the state's coal reserves. But opinions differ. When leaving the public reading room at the Division of State Lands after collecting the information about the last lease sale, a CEP researcher was buttonholed by Jack Kennedy who boasted, "Did you hear we brought in over half a million dollars for our beneficiaries in one day."

The bulk of New Mexico's state coal leases have been issued only recently, but already a large number have been bought and sold on the speculative market. Forty-three leases, or 20% of all valid leases, have been assigned to new leaseholders since their issuance. Like leasing in general, assignment activity is on the upswing. Twenty-seven of the 43 lease assignments have occurred in the last two years.

The 218 coal leases are distributed among 33 leaseholders. The five largest control 99 of the 218 leases and 48% of all leased state acreage—a level of concentration about average among the western states (see Table 6-3). The largest state leaseholder, Gulf Oil, owns 29 leases covering 13,199 acres. Gulf is also among the largest federal leaseholders, and through its subsidiary Pittsburgh and Midway, controls an 11,000-acre Navajo tract in New Mexico. Its New Mexico interests extend beyond coal, however: the company recently lost a lawsuit to United Nuclear, which accused it of attempting to monopolize New Mexico's uranium resources.

"We don't have any coal production on leased land. At least we don't know of any," Carlos Gallegos told CEP. CEP found, in fact, that no coal has ever been produced from any state coal lease in New Mexico. Neither Gallegos nor Kennedy were aware of any future mining plans which included state coal-bearing acreage.

If coal is ever extracted from state land, it will be subject to the highest royalty rates in the West. Leases issued in 1977 and after include a royalty rate

132

TABLE 6-3

TOP FIVE NEW MEXICO STATE LEASEHOLDERS

Leaseholder	Number of Leases	Acreage under Lease	% Total Acreage under Lease
Gulf Oil Corp.	29	13,199	12.3
Salt River Project	22	12,397	11.6
Cherokee & Pittsburg Coal & Mining Co.	18	10,554	9.9
Coastal States Energy	15	8,179	7.7
Hamilton Bros. Coal Co.	15	6,817	6.4
TOTAL	99	51,146	47.9

of 12.5% of the gross value of the coal or 86¢ per ton, whichever is higher. Leases issued in 1976 include a royalty of the higher of 8% or 55¢ per ton. Leases issued before 1976, however, include a miniscule 20¢ per ton royalty rate.

In the absence of production royalties, the Division of State Lands receives only advanced royalty and rental payments. For leases issued before 1975, rent was established at $25 per lease. Leases issued since then charge $1 per acre. In Fiscal 1977, the Division of State Land received $39,511 in rent payments—an average of just 37¢ per acre. Advanced royalty rates are substantially higher in New Mexico than in other western states, however. Lessees are required to pay $3 per acre for the first year of the lease, $4 per acre in the second year, and $5 per acre thereafter. Only Colorado assesses a higher advanced royalty rate—$10 per acre by the tenth year of the lease—but it waits until the sixth year of the lease to begin charging it. In Fiscal 1977, the state received $328,344 in advanced royalty payments.

As in other western states, rental payments and royalty and advanced royalty payments are distributed differently. Rent is turned over to the beneficiary institutions to become part of their operating budgets for the following year. Royalties and advanced royalties are placed in the State Permanent Fund and invested by the State Investment Council. Monies from all sources commingle in the Permanent Fund, but careful accounting records are maintained. Each year the interest earned from investments of the Permanent Fund is assigned to the institutions in direct proportion to their contribution to the principal (see Table 6-4). The State Investment Council, according to its Nineteenth Annual Report, seeks to conserve the fund's balance and "maximize the return to the beneficiaries."

The New Mexico State Permanent Fund's balance has grown to more than $736 million, by far the largest in the West. State lands are richly endowed with oil and gas reserves, the leasing of which has been bringing millions of

BENEFICIARIES OF NEW MEXICO STATE LANDS

Institution	Fund Balance
New Mexico State University	$ 3,336,312.48
New Mexico School for the Visually Handicapped	18,412,642.41
Charitable, Penal and Reform	5,562,630.00
Common School Permanent Fund	606,216,904.94
New Mexico School for the Deaf	18,541,197.14
Eastern New Mexico University	521,936.57
Improvement to Rio Grande	5,980,722.44
New Mexico State Hospital	3,114,424.38
New Mexico Military Institute	24,578,725.07
Miners Hospital of New Mexico	7,077,229.85
New Mexico Highlands University	369,399.73
Northern New Mexico State School	240,377.22
Penitentiary of New Mexico	15,727,946.82
New Mexico Boys School	101,523.43
University of New Mexico Saline Lands	53,129.14
New Mexico Institute of Mining and Technology	1,665,823.42
University of New Mexico	14,429,123.52
Western New Mexico University	373,853.72
Water Reservoir Permanent Fund	6,830,381.13
Public Buildings at the Capitol	3,372,587.75
TOTAL	**$736,506,871.16**

dollars into state coffers for decades. Each year, roughly $50 million is added to the principal of the fund; the State Investment Council expects the balance to exceed $1 billion within a few years. In Fiscal 1977, investment of the Permanent Fund yielded $44 million in interest income.

Since the amounts of money are so much greater in New Mexico than anywhere else in the West, it is here that the liabilities of the beneficiary institution structure are most blatant. Had the Permanent Fund's interest income been returned to taxpayers last year, each would have received $105. Instead, the money was allocated to the institutions according to the prescribed formula. The State Investment Council defends the rigid allocations as "important not only to the beneficiary institutions but also to the taxpayers in the state, since every additional dollar generated through investment of these monies is one less dollar that the taxpayer has to account for." CEP disagrees, finding it anachronistic that the New Mexico Military Institute, among others, benefits from development on coal-bearing land hundreds of miles away, while residents of coal rich counties must lobby for cash from the state to help ameliorate the adverse effect of such development on their lives.

CEP is further discouraged by the tremendous political leverage maintained by the managers of these funds. Jack Kennedy said that in New Mexico, where the funds are so large, "the Land Commissioner, in his own way, is even more powerful than the governor."

DEVELOPMENT

Among the many forecasts for New Mexico's coal future, the one proposed by the Western Coal Development Monitoring System, an ad hoc group of energy planners, is typical. The group predicts that the state's coal output will quadruple to 45 million tons a year by 1985. This forecast might be overly optimistic because it, like most others, is staked upon successful completion of the WESCO and CONPASO gasification projects. Much of the proposed coal development within New Mexico has been geared to meet the considerable needs of these plants: 1.8 billion tons of coal over 25 years. Utah International's Navajo mine, which could expand from its current output of seven million tons to 41 million tons a year, is slated to supply much of this coal. CONPASO has announced plans for two massive strip mines adjacent to its proposed complex, yielding another 25 million tons a year at capacity. Together, Utah International and CONPASO's mining plants account for more than half the capacity for all mining operations in New Mexico now on corporate drawing boards.

Should WESCO and CONPASO abandon their gasification schemes, only a limited number of in-state users of New Mexico coal would remain. Though planned expansion would be curbed in the absence of WESCO's large appetite, Utah International's Navajo mine will continue to supply coal to local power plants, foremost among them the Four Corners project. Western Coal Co.'s San Juan mine will serve the nearby San Juan power plant. San Juan production is expected to double to 2.4 million tons per year by 1978. Ideal Basic Industries plans on developing a smaller mine to feed sub-bituminous coal to its Tijeras, New Mexico cement plant.

Without gasification, the most important part of New Mexico's future coal economy will be the export market. Within 10 years, 10 or 15 medium-size coal mines in New Mexico will be exporting their products to electric utilities throughout the southcentral and southwestern states. Pittsburgh & Midway's McKinley mine, which supplies coal to Arizona Public Service Co.'s Cholla power plant in Joseph City, Arizona, is scheduled to quadruple production to five million tons per year by 1981. Peabody Coal Co. will operate the Star Lake mine it owns jointly with Cherokee and Pittsburgh Coal & Mining Co. to provide three million tons annually to the Coronado Electrical Generating Station in St. Johns, Arizona in the early 1980s. The Gramerco mine of the Carbon Coal Co. will begin producing 1.2 million tons per year in 1979 for Arizona Electrical Power Cooperative's Cochise, Arizona power plant. Tucson Gas & Electric Co. will begin production on its Gallo Wash mine in

San Juan County in 1983, reaching its planned 3.6 million ton capacity early in the 1990s. Coal from this mine will be used to supply three Springville, Arizona power plants. Coal emerging from Kaiser Steel's York Canyon and West York mines will be transported west to feed its steel mills in Fontana, California. And to the east, Texas Utilities will receive eight million tons of coal annually from the Hospah mine of its subsidiary, Jayco Energy Co. Production here should begin in the early 1980s. This mine is scheduled to grow into the state's second largest strip mining operation by the mid-1980s.

STATE POSTURE

New Mexico's attitude toward expanded coal development within its borders combines both pro-development and controlled growth impulses that have yet to blend to create a distinctive state posture. The traditional laissez-faire attitude of the state government has shifted somewhat in the 1970s with the creation of the Energy Resources Board, the Environmental Improvement Agency, and the Surface Coal Mining Commission—three agencies which spearhead a growing list of government organizations committed to alleviating and eliminating the adverse impact of coal development. In addition, the broad spectrum of environmental organizations, most notably the Southwest Information and Research Center and the New Mexico Citizens for Clean Air and Water, have been organizing growing citizen concern over energy development. Organizations including the Ship Rock Research Center and the National Indian Youth Council have emerged to deal with Indian energy development problems. Yet New Mexico's public interest groups and state government agencies generally lack the zeal characteristic of organizations in states like Montana and Colorado.

Carolyn Lindbergh of the New Mexico Department of Finance Administration told CEP that the statewide concern over coal development is limited by "the presence of so many serious needs here with many different origins. Coal development impact just does not seem so overwhelmingly important." Indeed, New Mexico must adapt a strategy to minimize the statewide impact of exponential and potentially catastrophic expansion of uranium mining and milling industries. Another concern is the federal government's plan to install the nation's largest respository of radioactive waste materials in the southern part of New Mexico. Large scale commercialization of coal conversion technology within state borders is a third major concern, sapping the energies of government and citizen groups alike.

"Planned coal development is not so extreme that the state needs to discourage or promote it" is an alternative explanation offered CEP by John Romero of the Department of Revenue. Still, it seems reasonable that the state at least develop the structure to monitor and, if necessary, regulate development.

136

If the Surface Mining Commission is any example, the state is ill-prepared for such a task. The Commission contains seven members, each a state official devoting only part-time work. The Commission's budget totals only $14,000 per year. It employs one part-time inspector. Since 1972 the Commission has reviewed seven permit applications filed for strip mining operations, though it has no paid technical staff to help it reach its decisions.

Kennedy, a Commission member, considers it "the model of western state reclamation." Yet the federal government has ordered New Mexico to upgrade its strip mining control program. The Commission is now debating whether to improve its program or abandon it, leaving the regulation of strip mining to the federal government. Kennedy is resistant to both options. He is a strong opponent of the new federal strip mining regulations. First, he claims, they are too strict. Second, he sees them as punishing the rest of the country for past environmental abuses in Appalachian coal fields: "Because of Appalachia, the federal government set standards that just don't make sense in the western climate." The bottom line of Kennedy's argument, however, is that "the new act is pure socialism."

Reluctance to permit the state government to intrude too far into the regulation of coal developers also informs New Mexico's mineral tax structure. Although the state has recently revised its severance tax and added an ad valorem property tax, the layered structure of mineral taxes still fails to reflect a strong commitment to controlling coal development.

New Mexico's mineral tax structure has three components. One which remained unchanged in the recent flurry of revisions is the Resource Excise Tax. It levies a tax of 0.75% of the value of the coal against either the mining company or the coal processing company but never both. Should coal be both mined and processed in New Mexico, it is taxed only once. Revenues from this tax totalled $632,000 in Fiscal 1977. Although this figure is expected to double by the end of the decade, it is still a small source of revenue to the state. All Resource Excise Taxes are deposited in the State General Fund.

The ad valorem mineral property tax, first assessed in 1976, constitutes the second layer of New Mexico's coal tax structure. Tax on the land's value and assets is levied against one-third of the property value. Private coal owners are subject to an additional surface value tax levied against one-third of the value of the coal it contains. Producers are assessed a mineral production tax on the coal's gross sale value less the cost of production. The rate of ad valorem tax varies from county to county.

The refurbished severance tax spotlighted the recent reform of New Mexico's tax structures. Until July 1, 1977, New Mexico assessed a severance tax against mineral operators of a mere 0.5% of the sales value of the coal at its first marketing point. Revenue from this tax collected in Fiscal 1977 was the highest ever, yet it totalled a mere $300,000. A new coal severance tax effective July 1, 1977, packs a much larger punch. The tax charges 38¢ per ton for steam coal and 18¢ per ton for metallurgical coal mined within the state. In addition, a surtax to be computed annually, will be assessed beginning July 1,

1978, to reflect changes in the Consumer Price Index. The New Mexico Department of Revenue estimates that the new coal severance tax will generate $4.2 million in its first year, 12 times more than the old severance tax generated the previous year. When the surtax escalator takes effect in Fiscal 1979, severance taxes are predicted to jump to $5.9 million.

Still, these figures are lower than in most other western states. John Romero of the Department of Revenue told CEP that legislative debate over the new coal severance tax seldom addressed its relatively low rates. Concern was focused instead on the effects of the revised severance tax on the state's active uranium development industry, which generates twice as much severance tax revenue as the coal industry. Moreover, the severance tax on coal contributes just a small fraction of the total severance taxes collected from all mineral industries within the state. In Fiscal 1977, revenues from all severance taxes—including copper, uranium, natural gas and coal—totalled $56 million; coal contributed just 7% of this total. The distribution of severance tax revenues reflects only a limited effort to meet the special problems mineral development creates. Of the $56 million collected in Fiscal 1977, $5.1 million were fed into the Severance Bonding Fund to retire debts incurred by grants to impacted communities for highway construction and general community assistance. The remaining $50.9 million however, was transferred to the State General Fund, bringing its balance to over $150 million. Investment earnings from this fund total $6.1 million and were added to the Severance Tax Income Fund's $1.7 million balance. The Income Fund then spent $5.1 million—less than 10% of the total severance tax collected that year—on capital outlays for affected communities.

Environmental activists and state government agencies clustered in Santa Fe are well aware of the adverse impact of rapid coal expansion. The new tax structure, however, suggests a confidence that coal development here will be regulated by internal mechanisms, as opposed to external ones like state taxes. It also reflects an attitude that the impact of coal development can be managed adequately without full use of the revenue generated by severance taxes. Most tax monies will help later generations of New Mexico's citizens, not those currently facing coal development. The new severance tax quietly bespeaks a commitment not to discourage coal development and not to encourage it, but merely to increase state revenues over the long-term.

7. NORTH DAKOTA

INTRODUCTION

Blazing the first officially-chronicled path across the American continent in 1804, explorers Lewis and Clark noted outcroppings of lignite coal in what was to become North Dakota. Using simple tools, the local Indians had been prying the coal from these outcroppings and burning it as a fuel. The explorers quickly learned also to use the area's coal resources for heating and blacksmithing. Homesteaders who followed appreciated the abundant fuel supply within easy reach. In 1883, the first commercial coal mine began operation in North Dakota, producing six tons dug out by hand. Such mines were soon distributed across the western portion of the state, and the mining industry has grown continuously since then.

The coal mining industry has always been a distant second to the state's predominantly agricultural base. For example, some 90% of the nation's durum wheat is grown in North Dakota. Any environmentally destructive mining activity is strongly contested by the powerful agricultural lobby. Nevertheless, coal production has increased steadily and orderly, though slowly, over the past century. In 1919 the state's first strip mine opened. By 1923, 226 North Dakota mines employed 2,145 workers—both figures endure as state records—and produced 1.4 million tons of coal. By the end of World War II, there were 111 strip mines producing 2.5 million tons. Throughout the 1940s, production continued to increase gradually as western railroads began purchasing North Dakota coal. At that time, however, oil was discovered in North Dakota and local homeowners turned from coal to cleaner burning fuels. Coal production in North Dakota declined from almost four million tons in 1951 to 2.3 million tons in 1958, the lowest production figure for the state since the depression.

The impact of the elimination of the home heating market on the coal industry in North Dakota was eased in the early 1960s by the construction of several coal-burning power plants within the state. Production again began to climb steadily, and by 1968 reached 4.4 million tons, a record at the time that has been surpassed each succeeding year. A steady growth rate hovering about 6% per year brought 1977 state coal output to 12.3 million tons.

North Dakota coal reserves lie in the Fort Union Formation, one of the largest coal deposits in the world, spanning eastern Montana and Wyoming as well as western North Dakota. While most of the coal in the western part of the Formation is bituminous or sub-bituminous in grade, all of North Dakota's coal is lignite, a lower quality coal. Lignite, while low in sulfur content, often contains more than 40% water. Its heating value is much lower than that of most other coals, ranging between 6,000 and 7,500 BTUs per pound.

What North Dakota coal lacks in heating quality it makes up for in sheer quantity. The state's recoverable lignite reserves are estimated at 16 billion tons, much more than the reserves of any other state and 50% of the lignite in the country.

There are 11 coal mines currently active in North Dakota. All of these use strip mining rather than deep mining techniques. The major companies comprising the mining industry within the state include Consolidation Coal, a Continental Oil subsidiary; Baukol-Noonan; Knife River Coal Mining Co., a subsidiary of Montana Dakota Utilities; and North American Coal Corp. Nearly all of the 12 million plus tons mined in 1977 were burned at power plants within the state. Only Knife River's Gascoyne mine supplies out-of-state customers—but even here the customer is the Big Stone power plant just across the South Dakota border.

The exclusive presence of lignite in North Dakota accounts for the state's coal marketing patterns. Exports have never been important to the coal industry in North Dakota. Because of its low heating quality, lignite is generally uneconomical to ship. Midwestern and southwestern utilities have found it cheaper to ship smaller quantities of higher grade Montana or Wyoming coal the extra distance than to ship North Dakota lignite to their power plants.

While the low heating value of lignite makes it unattractive for combustion to produce electricity and uneconomical to ship long distance, other characteristics make it attractive to the planners of the nation's first synthetic gaseous and liquid fuel plants. The high moisture content and non-caking character of lignite make it particularly suitable for use in coal gasification processes now nearing commercial application. American Natural Gas Co., Peabody Coal Co., and Tenneco Coal spearhead a list of energy corporations which are interested in bringing large scale gasification plants, and the mammoth strip mines to feed them, to North Dakota. Fifteen power companies have recently commissioned the Wentworth Brothers Co. to study the feasibility of converting North Dakota lignite into another synthetic fuel, liquid methanol or wood alcohol, which can be mixed with normal gasoline and used as a fuel for automobiles.

Coal conversion, whether producing synthetic natural gas or methanol from lignite, could mean a boom for North Dakota's coal fields. The Northern

Great Plains Resource Program estimated that the 1975 production of eight million tons is likely to jump to over 100 million tons by 1990 and could go as high as nearly 200 million tons if coal conversion projects proliferate.

As elsewhere in the West, implementation of coal conversion schemes has run up against serious obstacles in North Dakota, delaying commercialization and casting a cloud of uncertainty over the future. Most current problems reflect unresolved technical difficulties and uncertainties of the economic viability of a commercial conversion complex. But there is also strong local opposition, primarily by the state's agricultural interests, to the environmental impact of strip mining and the massive water requirements of coal conversion. As a result of this opposition, the North Dakota state government has created stiff taxes on coal development, strong land reclamation laws, a plethora of government agencies and task forces to develop and enforce new regulations, and a "go slow" policy that promises slow but steady coal development.

FEDERAL LEASING

The federal government plays a much smaller role in the North Dakota coal picture than in any other western state. Of the 22.4 million acres of coal-bearing land in North Dakota, the federal government owns only 5.6 million acres, or 25%, all in the western part of the state. In no other western state does the federal government own a smaller percentage of coal land.

There are fewer valid federal coal leases and less federal acreage under lease in North Dakota than in any other western state. Sixteen valid leases span 15,432 acres, less than 1% of the federal coal land within the state. Since 1973, five leases have been relinquished, causing a 25% decline in the number of federal acres under lease in the state. North Dakota is one of only two western states to experience an overall decline in leased acreage in the last five years. One new lease has been created—a 322-acre tract controlled by Consolidation Coal—by the "segregation" of an already existing lease into two smaller units.

Mining activity on federal land there actually declined in the past five years, making North Dakota the only state in the West to experience falling federal land coal output during this period. Federal land production in 1977 declined nearly one-third from 1973 levels to just 740,464 tons, only 2.6% of coal production from federal land in the West.

Federal output in North Dakota last year was staked on the production of four leases. No western state contained fewer producing leases. Only two mines there were of any significance, together accounting for more than 92% of the state's output. Knife River Coal Mining Co.'s Gascoyne strip mine in Bowman yielded 443,000 tons of federal coal. North American Coal Corp.'s Indian Head strip mine in Mercer County produced an additional one-quarter million tons.

Coal production on North Dakota's federal lands has been traditionally among the least significant in the West. Only in New Mexico has less federal coal been mined than North Dakota's estimated historical federal production of 30 million tons. The recent decline in interest in North Dakota federal coal does not signify the collapse of the entire traditional industry, however, but rather a readjustment to new, and lower coal demand predictions.

Still, the prospect of large gasification plants within North Dakota's borders creates expectations of coal development here greater than elsewhere in the West, though the development timetable stretches over several decades rather than years, as is the case elsewhere. The Department of Interior has projected that North Dakota federal land output will rise to 41 million tons a year—a 50-fold increase over 1977 production—if coal conversion plants become commercially viable.

If coal conversion interests move into North Dakota in a big way, they must face extensive land use conflicts on federal land. All federal coal land leased to date is privately owned at the surface. The government has leased only the subsurface mineral rights. Vigorous efforts by local ranchers and vocal conservationists virtually ensure that some surface owners will not willingly part with their property to accommodate strip mine developers. In no other western state are surface–subsurface conflicts so extensive. Consequently, resolution of the surface–subsurface rights for any new coal development proposal on federal lands will be neither swift nor pleasant.

STATE LEASING

Upon North Dakota's admission to the Union in 1889, 2.5 million acres were transferred from the federal government to the new state government. Federal legislation at the turn of the century turned over additional acreage to state control. By 1977, the state of North Dakota had amassed control of roughly 3.19 million acres of land. The state is thought to be the second largest coal owner in North Dakota—behind the federal government—with 900,000 acres of lignite-bearing land.

Throughout most of the century, state land in North Dakota has been informally transferred to coal mining interests for development. "In the early days of statehood," explained Richard Lommen, Director of the North Dakota Board of University and School Lands, "we just gave the land away. In the '20s and '30s, we allowed people to mine coal free of charge." Nearly all the lignite mined from state land during this period was used by local farmers for home heating. "It was the Depression," added Lommen. "Farming was so important that we took our state coal for granted."

In the late 1950s, the North Dakota state legislature enacted legislation and promulgated regulations to structure a state leasing program. At this time the Boards of Land Commissioners and University and School Lands were established to administer all coal leasing activity. Since then, 10 leases spanning 3,838 acres have been issued, making North Dakota's the smallest state leasing program in the West (see Table 7-1).

TABLE 7-1

NORTH DAKOTA STATE LEASE ISSUANCE

Years	Number of Leases Issued
1959	1
1960-1967	0
1968	1
1969	2
1970	1
1971	1
1972-1976	0
1977	4

From 1959 until 1971, the program accounted for the competitive issuance of six leases. The first was awarded in November 1959 to Knife River Coal Co. without requiring payment of any cash bonus. Nine years later, a second lease was issued to Baukol-Noonan for its bid of $1 per acre. The four leases issued from 1969 to 1971 drew winning bids ranging from $1 to a high of $2.10 per acre. Only two of the six competitively leased tracts attracted more than one bidder. The average winning bonus bid for the 3,118 acres of competitively leased state coal land was $1.16 per acre, the lowest encountered by CEP among the five states conducting competitive leasing at any time in the history of their programs.

Lease terms were correspondingly lenient. Royalty rates were established at 10¢ or 15¢ per ton. Rent was just $1 per acre. The first lease was issued for a 20-year term; the other five were valid for 10 years. Each lessee possessed an unlimited right to renew its lease.

Lommen admits that this leasing program was unregulated and maladministered, mirroring the early leasing efforts of other western states: "They just came in and made an offer and we gave the lease away." Unlike other western states, however, North Dakota gave away only a very few leases under this inadequate program. Before losing control of even more state acreage, the Board of Land Commissioners instituted a moratorium on the issuance of new coal leases in 1971. Its primary concern was neither lease speculation nor low revenues, but the effects of strip mining on prime agricultural land in the western part of the state. State agencies worked throughout the moratorium to draft new reclamation legislation to better protect farmland. "We rank right up there now," asserted Lommen, with reclamation requirements among the most exacting in the nation.

On June 26, 1975, the Board of Land Commissioners promulgated new leasing rules and regulations. The new procedures replace competitive leasing with a program of negotiation—the only such state lease issuance program in

143

the West. Less than two years later—before any new leases had in fact been negotiated—"interim leasing criteria" were adopted to further reform the old program. Together, the new regulations and criteria afford the Board of University and School Lands sufficient latitude to regulate coal development with a precision and direction unique among western states.

The new lease application process should discourage all but the hungriest of developers. Prospective lessees must submit a proposed lease package, including cash bonus, rent, and royalty payments, to the state. Applicants must convince the Board that the state land is necessary for the continued operation of an ongoing mine or as a reserve for a mine that will begin production within five years. They must commit themselves to projected production dates for each state tract and prove that such a plan constitutes a logical pattern of coal development. They must further prove they have customers for the coal to be mined from the tract.

Prospective lessees who successfully document their intention to mine then sit down with Board officials to hammer out precise lease terms. These too will substantially raise the ante for obtaining a coal lease. Minimum rental payments are to be $1 per acre for the first five years of the lease, $1.50 for the sixth and seventh years, and $3 per acre thereafter. Royalties may not fall below 50¢ per ton or 6% of the value of the coal, whichever is greater. Lease terms endure for 10 years. No lease may be renewed unless "mining operations are being conducted without unreasonable cessation." Lessees may not renew their inactive coal leases merely by making advanced royalty payments.

If the Board accepts the applicant's proposal, the lease terms and a summary report prepared by the Land Commissioner are circulated to 22 state agencies for a 90-day review period. If no compelling objections result, the Land Commissioner may authorize a competitive lease sale. At the auction, any interested party may best the negotiated terms and obtain the lease for itself. After the sale, a public hearing is held at which testimony concerning the proposed lease is solicited. The Land Commissioner, considers this public testimony, weighs counter offers, and ultimately decides whether or not to issue the lease. Even after the original negotiations and lease sale, the Land Commissioner is empowered to require additional bonus, rental, or royalty payments or refuse to issue the lease.

"We are proceeding at a cautious pace with implementing the short-term criteria and the new leasing regulations," says Lommen. Only four leases covering 720 acres have been issued under the new program. Each was issued on December 21, 1977. Three were awarded to Baukol-Noonan and one to Consolidation Coal, both companies are already involved in mining North Dakota's lignite reserves. The modest terms of all four negotiated leases were virtually identical.

The negotiations brought in only minimal cash bonus payments. Each company paid just $1 per acre—even less than earlier competitive sales had generated. The issuance of the four leases brought the state $720, and raised to

TABLE 7-2

REVENUE FROM NORTH DAKOTA STATE LEASE SALES

Method of Leasing	Number of Leases	Total Revenue Collected
Competition	6	$3,637
Negotiation	4	720
TOTAL	10	$4,357

$4,357 the revenue collected by the state from the issuance of the 10 currently valid leases (see Table 7-2).

Better results were achieved in other areas of negotiation. Rent for each of the four tracts will rise from $1 per acre for the first five years to $3 per acre for years six through eight and level off at $5 per acre thereafter. Royalty rates of 10% of the value of the coal, minus severance tax payments, will be assessed. The Land Board estimates that the effective royalty rate will average 38¢ per ton.

A larger advantage of the new leasing procedures is that they will, in Lommen's words, "definitely eliminate all speculation." Though never as widespread as in other western states, speculative assignments of state leases have occurred in North Dakota. After Thomas Pearce and Lenus Volk bested Husky Industries' bid of $1 per acre for a 300-acre lease tract in 1970, they turned right around and assigned the lease over to Husky Industries in 1976. Michael Speaks won rights to a 298-acre lease in 1969 by bidding $1.26 per acre, and one year later assigned the lease to North American Coal, the other bidder at the original lease sale. Requiring guaranteed coal production from an approved mining plan, the new leasing program promises to make such assignments a relic of the past.

All four of the newly-issued leases will be in production shortly. The three issued to Baukol-Noonan adjoin the company's Larsen mine, one of the oldest coal mines in the state. Most of the lignite produced there is sold to American Crystal Sugar Co.'s sugar beet factories in North Dakota and Minnesota. Smaller amounts are sold to North Dakota University, which consumes roughly 88,000 tons of lignite each year. Consolidation Coal Co. will expand its Glenharold mine onto the 320-acre tract it recently acquired by negotiation. This Continental Oil subsidiary expects to mine 300,000 tons of state reserves and market them to Basin Electric Co.'s Leland Olds plant.

As the only western state where firm plans to systematically expand coal production on state land exist, North Dakota can be counted on to dominate western state coal output in the years to come. Already in 1977 its five active leases produced 1,267,893 tons of coal, second only to Montana in the West. All of Montana's production, however, emerged from one lease tract—part of the mammoth Decker mine—which will be mined through shortly. On the other hand, North Dakota, with five of its ten valid leases now active, has by

145

far the highest percentage of active leases in the West. In fact, in no other western state did more than 2% of the valid leases produce coal in 1977.

Two of the currently producing state leases in North Dakota are controlled by North American Coal Co. Their combined output was 480,000 tons in 1977. Knife River Coal Co.'s lease, part of the Beulah mine, yielded more than 285,000 tons in 1977. Husky Industries produced 23,000 tons of state-owned lignite to be converted into charcoal briquettes and sold commercially. Baukol-Noonan's Center mine stretches onto a parcel of state land which generated roughly 475,000 tons last year.

Current revenues accruing to the state from coal leasing activity remain minimal. Since every valid lease is now paying $1 per acre in rent, the state collected $3,838 in rental payments last year. Production royalties totalled $132,263, the second highest royalty collection among the western states, but averaged only 10.4¢ per ton, the lowest in the West. The average rate and total collection of production royalties will increase in the near future, however. The terms of older leases, which included extremely low rates, will be evaluated upward upon their expiration. In March 1978 the state's oldest lease completed its initial 20-year term and came up for renewal. Instead of rubber-stamping the old contract, as many western state leasing agencies continue to do, the Board of University and School Lands upped the royalty rate from 10¢ to 35¢ per ton. The 10-year terms of the other five leases issued under the old competitive program will expire and come up for readjustment within the next three years. Royalty rates and collections will also increase substantially when the new leases, which require royalty payments of 10% of the value of the coal, begin production.

Fourteen Congressionally-selected institutions benefit from minerals activity on state land (see Table 7-3). Rent payments are turned over directly to the institutions and incorporated into their operating budgets for the following year. Royalties are placed in the state's Permanent Fund and invested. Investment earnings are then turned over to the institutions in proportion to their contribution to the Fund's principal. The current balance of the North Dakota Permanent Fund is $80.6 million, making it the fourth largest Permanent Fund among the six western states.

Looking towards the future, Commissioner Lommen's aspirations for state coal leasing in North Dakota are exemplary. He is anxious to expand the leasing program. "We certainly don't believe that our coal should be left in the ground," he told CEP. "In fact, we would be remiss not to give our coal to developers." However, the Board will only lease state land needed to complete logical mining units predominated by federal or private land tracts, and only if it can exert control over the type, pace and environmental effects of development. Unlike other western states surveyed by CEP, North Dakota's leasing officials are content to proceed slowly. "We're looking for sustained, rather than immediate income," summarizes Lommen. "We want to maximize values, not money."

TABLE 7-3

BENEFICIARIES OF NORTH DAKOTA STATE LANDS

Institution	Acreage	Fund Balance
Common Schools	2,523,383.78	$67,362,793.81
North Dakota State University	129,999.98	2,989,567.31
School for Blind	30,025.76	690,034.10
Capitol Building	82,326.14	891,045.48
School for Deaf	39,966.53	811,232.22
State Hospital	20,000.98	517,519.35
Ellendale College	39,997.24	779,372.64
Valley City College	50,003.69	991,420.63
Mayville College	30,002.21	614,481.23
Industrial School	40,024.94	912,761.02
Science School	39,997.16	720,116.85
School of Mines	40,003.51	911,930.02
Soldiers Home	39,972.36	773,219.74
University of North Dakota	86,066.00	1,714,971.37
TOTALS	3,191,770.28	$80,680,465.77

DEVELOPMENT

Coal development in North Dakota is concentrated in Oliver, Mercer, Ward and Burke Counties in the western part of the state. Each of the 11 currently active coal mines in the state are located in one of these counties, and each employs strip mining techniques. The largest is Consolidation Coal Co.'s Glenharold mine. Partially situated on a federal lease in Mercer County, it produced 3.4 million tons, or about one-quarter of the state's entire coal output last year. Consolidation Coal hopes to expand production soon to just under four million tons per year, and to market its lignite to the nearby Leland Olds #1 and #2 power plants, operated by Basin Electric Co.

Baukol-Noonan, Inc. operates the Center mine in Oliver County, the state's second largest mine, which generated 2.6 million tons of coal in 1977. The company plans to more than double this mine's capacity to 4.4 million tons per year in order to supply low sulfur lignite to the new Square Butte power plant. The nearby Beulah mine operated by Knife River Coal in Oliver and Mercer Counties yielded another 1.9 million tons in 1977. Knife River seeks to triple production capacity there to over four million tons per year by 1982 to meet the needs of the proposed Coyote power plant.

Consolidation Coal and Baukol-Noonan operate smaller active mines in Ward and Burke Counties, respectively, that have been contributing to North

Dakota's state coal output for more than 40 years; last year they combined to contribute 0.75 million tons of coal. Finally, North American Coal Corp. operates a small mine in Mercer County, which produced about 0.78 million tons last year.

Expansions and reopenings of existing mines that have dominated state output throughout the last decade can be expected to increase North Dakota output to over 20 million tons by the mid-1980s. Nearly all of this coal will be burned in neighboring "mine-mouth" power plants. Several new mines slated for initial production within the next five years will also serve local power plant units. Among the most significant of these planned mine openings is the Falkirk Key mine in MacLean, operated by the Falkirk Mining Co., a subsidiary of North American Coal Corp. Production there could reach as high as six million tons per year by 1981. The lignite, which will be strip mined, will supply the nearby Coal Creek power plant complex.

Gasification proposals are by far the largest and, some say, most menacing specter on the horizon of coal development in North Dakota. The American Natural Gas Co. has developed plans for a $900 million gasification plant comprised of four units requiring as much as 52 million tons of lignite annually by the 1990s. This one plant would consume, in one year, more than four times the entire state's coal output in 1977.

To date, the company has spent more than $20 million developing plans for the complex. American Natural Gas Co. representatives announced in early 1978 that they will need federally-guaranteed funding of at least $675 million by October 1, 1978, if they are to go ahead with their project.

Reluctance to subsidize large commercial gasification projects led federal officials to suggest instead that American Natural Gas seek out partners in its enterprise to defray the costs. People's Natural Gas Co. of Chicago has already been brought in as such a partner. Financing problems remained unsolved, however. In an early June 1978 breakthrough, American Natural Gas acquired three additional partners: Columbia Gas System, Inc., Tenneco and Transco. The five companies have enlisted the support of the Department of Energy in a new demand for federal guarantees. They have asked that they be able to recoup their losses, should the project ultimately be scrapped, by passing special rate increases on their current customers. This petition still awaits the approval of the Federal Energy Regulatory Commission. Even if they get the go-ahead, construction would not begin until early 1980 and the expected date of initial operation would be re-targeted for the late 1980s.

Already North Dakota mining companies have revealed plans of enormous scope to meet the mammoth coal requirements that the American National Gas gasification plant would require. The Couteau Properties Co. has announced its intention to construct and operate a strip mine in Mercer County that would yield 14 million tons annually by 1984. AMAX has plans for another 14 million ton per year strip mine to feed the plants. AMAX's mine is to be in Dunn County and is expected to reach capacity sooner than that of the Couteau Properties.

There are several other gasification projects on the drawing boards in North Dakota, though none are as far along as the American Natural Gas proposal. Peabody Coal hopes to build one gasification complex in the state, to be fed by coal mined from the company's holdings in Grant County. Tenneco Coal Co. controls one billion tons of strippable coal in Golden Valley County where it too hopes ultimately to operate a large mine to feed a nearby gasification plant it intends to construct.

Opposition to gasification proposals by North Dakota's agricultural interests, conservationists and ranching industry, runs deep. Furthermore, the federal government does not appear eager to subsidize the nascent industry. Should the gasification schemes thus be stalemated, synthetic fuel advocates have another coal conversion program in mind: the production of liquid methanol fuel from coal. Early in 1978, 15 power companies commissioned the Wentworth Brothers Co. to study the feasibility of a $1.5 billion methanol plant in North Dakota to produce 7.5 million gallons of wood alcohol daily. Critics cite methanol's high initial costs of $3 per million BTUs, roughly equivalent to that of propane, as prohibitive. Furthermore, the environmental problems associated with a large scale methanol production industry equal those of gasification. Others, however, indicate that power plants can burn methanol cleanly in a turbine without pollution control equipment, making the true cost and environmental impact of methanol burning actually lower than that of coal-fired boilers. The study is to be completed by the end of 1978.

Abuzz with the prospect of a gasification-stimulated boom, observers of the North Dakota coal industry are debating the actual pace of future coal development. Coal conversion proponents hope to utilize western North Dakota's vast lignite reserves quickly and create a boom within the next decade. Others—perhaps those who know the area and its inhabitants better— like State Engineer Vern Fahey of the North Dakota Water Commission, see development at a standstill with no significant change in coal output through the year 2000. CEP found that most state officials defer to the forecast of expert John Smith of the North Dakota Lignite Council, who anticipates continued growth at a steady, orderly pace.

STATE POSTURE

North Dakota probably has a stronger agricultural base than any western state slated for large scale development. It also has a long history of fervent opposition to outside interference by large eastern industrials. Populist and socialist organizations such as the Nonpartisan League, the Farmers Union, and the Dakota Farmer Alliance proliferated early this century, and several remain very active in state and local government. Many local farmers have brought their political clout to the struggle against large scale coal expansion now waged by environmentalists, conservationists and traditionalists within North

Dakota. Together they have found a sensitive ear in the state capitol, provoked the state's growing resistance to gasification, and helped shape its concern over the impact of coal expansion.

The North Dakota Water Commission has often refused to grant permits to coal-related projects requiring significant diversion of waters which could be used for agriculture. The Public Service Commission has administered reclamation standards which, since their promulgation in 1969, have been redefined and upgraded in each legislative session so that they now rank among the most exacting in the nation—though several serious loopholes remain. The State Land Office mandated a six-year moratorium on the issuance of state coal leases which ended in 1977. Mercer County, the seat of coal development, is now the site of a Department of Energy-funded Energy Development Board established to plan coal development and monitor energy flow. This project is one of only two such federally-commissioned experiments in the country. The North Dakota Regional Environmental Assessment Program (REAP) is now in its third year of developing economic and demographic models to help the state cope with coal development.

The recent strengthening of the coal severance tax reinforces North Dakota's efforts to limit coal development expansion and its adverse effects. Until July 1977, the state taxed coal miners at a flat rate of 50¢ per ton with a 1¢ escalation for every 3% increase in the Wholesale Price Index (WPI). In Fiscal 1977, the last year the old structure was in effect, roughly $6 million was collected in severance taxes. The new severance tax schedule authorizes an increase in the base tax rate to 65¢ a ton with a 1¢ increase for every 1% rise in the WPI. The tax is never to fall below 65¢ a ton, even if the price index dips. Adjustments to the severance tax rates will be made quarterly. Severance tax yield is projected to jump to $8.2 million in Fiscal 1978 and $12.2 million in Fiscal 1979. Already the Escalator Clause has increased the tax rate to 66¢ per ton and is expected to increase it to 69¢ per ton in the second and third quarters of Fiscal 1978.

Along with the new severance tax schedule, the 1977 legislature authorized a new pattern of distribution of severance tax revenues. Foremost among the changes was the establishment of a Coal Development Impact Office (CDIO), to receive 35% of all monies generated by the tax. The CDIO will administer grants to affected cities, counties and school districts to help meet the extraordinary impact of coal development and attendant population growth. Already the Office has approved $2.1 million in grants, according to Director Ralph Dewing, mostly for capital improvements such as upgrading waste disposal systems and water treatment plants in small cities with populations under 2,000. Apart from offering cash for capital improvements, CDIO seeks to upgrade services in its target regions. Law enforcement and fire protection agencies are scheduled to receive high priority funding within the next two years.

The 1977 legislature allocated 15% of the severance tax revenues to the Board of University and School Lands to be placed in a nondepletable Natural

Resources Trust Fund. The Fund now totals $1 million. From it, the Board is authorized to issue loans to development-impacted regions whose needs are corraborated by the CDIO. Although as of January 1978 no loans had been issued, two inquiries were pending—one from a city seeking to refurbish its city hall, and one from a school district for school construction purposes.

Counties from which coal is mined are allocated 20% of the severance tax revenues to be distributed proportionately according to the actual tonnage severed. Cities within the producing counties are then to receive 30% of the county allocation in proportion based on their population. The County General Fund will receive 40% for general government purposes and the remaining 30% of each county share is to be distributed to the county school system.

The final recipient of the severance tax revenues is to be the State General Fund. Its share is to be 30% of all revenues collected. In addition, all interests from loans issued by the Board of University and School Lands will accrue to the General Fund.

Unlike other western states, there is no ad valorem tax on coal development in North Dakota. State officials believe that periodic boosts in the base level of the severance tax schedule complementing the quarterly rate adjustments will discourage uncontrolled development while providing the cash to meet any adverse impact. The only other coal-related tax in the state is a coal conversion levy, which applies to power plants larger than 120 Mw, and gasification and liquefaction plants using more than 500,000 tons of coal per year. Revenues from this tax are allocated to the General Fund (65%) and the specific counties where the power or conversion plants are sited (35%). Coal conversion revenues collected in Fiscal 1977 totalled just under $1 million, slightly less than the amount received the prior fiscal year. Although revenues from conversion are now small, they would jump substantially if gasification ever becomes a major industry in the state.

The efforts made by the North Dakota government to limit the adverse impact of coal development while generating revenue to government treasuries around the state exceed those of every other state examined, with the possible exception of Montana. Such efforts as these, though still too limited in the eyes of many environmentalists, best express the aspirations of North Dakota's farmers, conservationists and cowboys who have led their government in seeking to protect their land and their lifestyles.

8. UTAH

INTRODUCTION

In 1869 when the silver sledge drove the golden spike, creating America's first transcontinental railroad, prospectors were already struggling to uncover the riches that lay beneath the mountains of Utah. The railroads created new markets and brought out even more miners, so that by the turn of the century hundreds of active mines sprinkled the new state.

Current coal miners are hopeful that the world-wide energy crisis and the nation's turn toward coal will again stimulate the Utah coal industry, but they are worried. The very qualities that made Utah coal development so attractive decades ago now limit future development there. Utah's coal lies deep underground, an advantage before the advent of strip mining technology, but now a liability. "There will be no coal boom in Utah," according to Jackson Moffit, Area Mining Supervisor for the U.S. Geological Survey in Salt Lake City, "because our coal just cannot be strip mined. When transportation and freight costs to the major midwestern markets are figured into production costs, we can't compete on a BTU basis with Wyoming." Unlike Montana's Burlington Northern Railroad, Utah's transcontinental line, once the pride of America, has deteriorated badly.

Recent changes in the coal market, however, could prove projections such as Moffit's incorrect. While it is true that Utah's coal industry has suffered because of its distance from midwestern markets, it may no longer have to rely solely on markets to the east. California's ever-expanding need for new energy sources promises to encourage development of Utah's coal fields. In the past, coal has not been burned extensively in California because of strict air pollution codes, but with nuclear power hamstrung, and alternative technology slow to emerge, utilities are turning to coal.

Utah has more than enough coal to satisfy the needs of California utilities for decades. Though many western states contain more total coal reserves, Utah is richly endowed with the high quality, low sulfur, bituminous grade, that is most attractive to developers. Its 23 billion tons of bituminous resources are second only to Colorado's in total quantity, but Utah's coal is generally more accessible then Colorado's.

Two coal regions have traditionally been the centers of Utah's coal production. Most of the 359 million tons of coal produced in the state's history have come from Carbon, Emery, Sevier, and Grand Counties in the central part of the state. The Book Cliffs field in these counties is still the most productive area in the state. The Wasatch Plateau, also in central Utah, is another important source of coal, providing fuel for the huge Huntington Canyon generatng plant. South of the Wasatch Plateau is the Emery Field. Without access to sophisticated transportation systems, the two active mines there have been dispatching their small coal output on trucks to Price and Salina for distribution.

The second major coal rich region in Utah lies in the southern part of the state, spanning Kane, Garfield and Wayne Counties. There, the Kaiparowits Plateau contains the largest untapped mineral reserve in Utah, including four billion tons of recoverable coal reserves. The coal there is generally of lower quality than that found in central Utah, but proximity to southern California markets and Lake Powell, Utah's major water source, enhances its attractiveness. No major coal mines have begun to operate, but the plateau is slated for large scale coal development in the next decade. Such plans have provoked considerable opposition because of Kaiparowits' proximity to the "Golden Circle" of national parks, including Grand Canyon, Zion, Bryce Canyon, Canyonlands and Capitol Reef. Prospects for development of other coal fields in southern Utah vary. West of Kaiparowits, the Alton field contains strip coal which, a recent University of California at Los Angeles study concluded, could be effectively marketed in California. Development on coal fields just north of Kaiparowits will be discouraged by an uncertain water supply, lack of adequate transportation facilities, the low quality coal contained there, and the geographic remoteness of the area.

Large quantities of high quality coal nearby are not all that California utilities find appealing in Utah. The state's arid terrain has attracted fewer ranchers and farmers than most other western states, so there are fewer competing interests for public land and fewer potential conflicts involved in development projects. Both the state and federal governments, which control virtually all coal reserves in Utah, have traditionally been kind to miners. While energy developers around the country face a waft of bureaucratic red tape, Utah relishes its future role as energy supplier.

FEDERAL LEASING

Federal coal has featured more prominently in the history of Utah coal development than in any other western state. This pattern should continue in the future. Fully 70% of the land within the state is managed by the federal government, including more than 80% of the coal land. More federal coal leases—198 in total—have been issued in Utah than in any other state. These leases cover 274,275 acres, nearly one-third of all the federal coal-bearing acreage under lease in the six states examined by CEP. The Department of Interior estimates that 4.2 billion tons of Utah coal have been leased by the federal government—about one-fourth of all leased federal coal in the West. An additional 1.2 billion tons of federal reserves are buried under the 138,000 federal acres in Utah currently under preference right lease application.

Since coal production on federal land began more than 50 years ago, about 95 million tons of public coal have been extracted, roughly 50% more than has been mined from federal land in any other western state. In the past five years, annual coal production from federal land has increased 96% to 5.8 million tons in 1977. Roughly 65% of Utah's current coal production comes from federal land. Fourteen of the 17 active coal mines in the state operate, at least in part, on federal domain.

Because federal coal so dominates in Utah, the uncertainties in federal leasing priorities during the moratorium were felt more strongly there. "Coal leasing has been dead since 1970," Supervisor Moffit told CEP. Only two new leases have been issued since 1973. Coal companies are "feeling a real bind. They need additional resources, new lands to mine."

Nevertheless, CEP found that recent coal development activity on federal land in Utah has been lively. Coal production on federal land has nearly doubled since 1973. The number of active coal leases increased by 50% to 21 in 1977. That these 21 leases represent only 11% of the total outstanding coal leases signifies that there is plenty of room for expansion of coal production on federal land in Utah even without issuing new leases.

Another indication of activity on federal coal land since the moratorium is the prevalence of "lease assignments." CEP found that between 1973 and 1977 the ownership of 40 of the 198 coal leases changed hands. Moffit criticized the federal government for encouraging this speculation by "giving away all of Utah's land and then closing the market." Although he believes that the assignment boom will end as leases eventually devolve into the hands of companies with the capacity and intent to mine coal, Moffit acknowledges that "the speculators are still selling among themselves. Those who have leases are asking exorbitant prices for them. To undo this, the key is to open up leasing again."

The need for renewed leasing has been challenged by environmentalists, who argue that much of the 4.2 billion tons of coal currently under lease is still not committed to production. They further fear that if the 38 currently pending preference right lease applications are converted into leases, they will create an additional burden of public coal transferred to the private sphere without regard to resource or land use planning or environmental protection. Renewed leasing might also allow the coal industry to make inroads into Utah's pristine area of wilderness and national parks.

Proponents of new leasing counter that since most Utah coal requires deep mining procedures, minimal environmental disruption would occur to land surfaces now used for grazing or recreation. The federal government would, therefore, be able to implement multiple land use practices on public land, a major directive of the Federal Land Policy and Management Act of 1976. They further contend that Utah's fortuitous shortage of conflicts between surface and subsurface land owners testifies to its unique suitability for renewed leasing. Since only 5% of federal coal-bearing land in Utah has been homesteaded, the federal government retains ownership of the surface as well as subsurface mineral rights on most Utah land. Often, then, surface and subsurface development rights can be leased simultaneously. By contrast, 97% of leased federal coal land in Montana is privately owned at the surface.

The Department of Interior expects that private developers will flock to Utah's federal land, if they are given the chance through resumed leasing. Its Final Environmental Impact Statement on the Proposed Federal Coal Leasing Program predicts that by 1990, Utah's federal coal production will jump up from its 1974 level by a factor of 12, while federal production in the West at large will only triple.

STATE LEASING

When Utah entered the Union in 1896, Congress was particularly generous, granting the new state government control over four sections—2, 16, 32, and 36—in every surveyed township. The state government now controls 8% of the total land area within Utah. Most of this state-owned acreage exists in scattered, 640-acre parcels, although some larger consolidated land plots exist in the Book Cliffs area.

Geologists have determined that roughly 574,000 of the 3.6 million acres owned by the state bear coal. Since 1942 the Utah Division of State Land has leased to individuals and corporations the coal development rights on state-owned land. Five hundred and fourteen leases spanning 543,557 acres have been issued. Most are between 500 and 1,000 acres in size, reflecting the checkerboarding of state-owned sections (see Table 8-1). However, CEP discovered nine leases larger than the statutory limit of 2,560 acres. Fully 74% of all leases have been issued in the last five years when the federal leasing moratorium forced energy developers and land speculators to turn to state-owned land (see Table 8-2).

Utah has welcomed prospective lessees into its leasing program with the most lenient, inadequate, and maladministered state leasing program in the West—with the possible exception of Wyoming. Nearly all of the leases have been issued for nominal filing fees, without regard for land or resource use planning. Rent and royalty requirements are extremely low, even when measured against the low rates of other western states. No strides have been taken to insure orderly coal development or protection of the environment.

TABLE 8-1

UTAH STATE LEASE ACREAGE

Acreage	Number of Leases
0-249	55
250-499	53
500-749	188
750-999	17
1,000-1,499	66
1,500-1,999	48
2,000-2,499	42
2,500-	45

TABLE 8-2

UTAH STATE LEASE ISSUANCE

Years	Number of Leases Issued
1942-1951	4
1952-1961	9
1962-1963	43
1964-1965	16
1966-1967	13
1968-1969	17
1970-1971	27
1972-1973	43
1974	85
1975	137
1976	39
1977	81

"We are more liberal than any other state in granting leases," claimed Donald Prince, Assistant Director of the Division of State Land. State law requires the Division to issue a lease to any interested person, excluding state employees, who are "not in default of the laws of the state" or to any corporation "qualified to do business within the state." Applicants need not demonstrate any desire or capacity to mine coal, nor must they even prove the existence or suspected existence of recoverable coal reserves. Four hundred seventy-one coal leases representing 92% of all those issued to date have been awarded according to open land procedures for only a $5 filing fee. Another

TABLE 8-3

REVENUE FROM UTAH STATE LEASE SALES

Method of Leasing	Number of Leases	Total Revenue Collected
Open Land	471	$2,355
Segregation	20	140
Competitive Bidding	23	31,527
TOTAL	514	$34,022

20 leases have been created by segregation, netting the state a total of $140. The competitive issuance of 23 leases covering 15,923 acres has brought the state $31,527 in cash bonus payments (see Table 8-3).

The Division of State Land holds competitive lease sales only when an existing lease becomes reavailable for leasing due to relinquishment or termination. When this occurs, the state advertises the tract's availability and extends interested applicants a 15-day simultaneous filing period to submit sealed bids. At the end of the 15-day period, the bids are opened and the highest bidder is offered the lease. The state is empowered to reject only bids falling below 50¢ per acre.

Winning bids at the special lease sales held to date average only $1.98 per acre. Only one bidder submitted an offer for 57% of these leases, bidding an average of $1.65 per acre. Thirty-five percent of the lease sales attracted two bidders. No lease sale ever attracted more than three. No bid ever exceeded $12.20 per acre.

The cost of maintaining a lease is as nominal as the cost obtaining one. Lessees are required to pay rents of 50¢ per acre for the first two years of their lease's terms and $1 per acre thereafter. Although lease terms are to expire after 10 years if active mining is not occurring, Utah permits its lessees to renew their contracts indefinitely simply by paying advanced royalties of $1 per acre per year thereafter. Since advanced royalties are credited against rent, this is basically only a bookkeeping change—with one slight twist. Advanced royalties, unlike rent, accumulate as credit against any future production royalties.

Such low lease terms encourage leaseholders to keep their leases even if they harbor no mining intentions or expertise. State law permits leaseholders to barter their leases for cash on the assignment market without reporting the size of the bonus payment to the Division of State Lands. They must merely inform the Division of the ownership change and the overriding royalty rate to be paid to the assignor by the assignee if production occurs. The state may cancel the assignment royalty if it determines that the overriding royalty is so high as to ultimately discourage production. It has never exercised this power. Instead, assignments have occurred at a rate unmatched in the West. CEP found that 61%, or 315, of the 514 leases were plucked by their current owners off the assignment market. Since virtually all the state's coal land has already been leased, developers and speculators must now look to the assignment market to enter Utah's state leasing program. Recently, they have been doing so in increasing numbers: in 1976, for example, 120 lease assignments occurred while only 35 new leases were issued (see Table 8-4). Since the state receives only a small filing fee from these assignments, Prince told CEP that "the state really should exercise more control over private business deals involving the state's resources." But he noted that "there is no pressure for change from anyone. And if we attempted to reform the provision, there would be a lot of opposition."

High risk energy speculators along with large mining companies dominate the list of Utah's largest leaseholders (see Table 8-5). The largest leaseholder, Coastal States Energy of Houston, controls 82,000 acres of Utah state land along with its extensive federal holdings. John Adams and Wilene Shumway—the third and fourth largest leaseholders in the state—own most of their leases jointly. For example, 34 of the 56 leases Shumway controls are 50-50 joint ventures with Adams. Another 18 are shared with Adams and L. Laughlin, the eighth largest leaseholder in the state.

None of the top five leaseholders has ever produced a single ton of state coal. Last year, only seven of 514 leases were actively producing, contributing to the outputs of four different mines. An eighth lease, controlled by 5M, Inc. will soon come into production as part of the John Henry mine. In 1977 four

TABLE 8-4

UTAH STATE LEASE ASSIGNMENTS

Years	Number of Assignments
1951-1957	1
1958-1964	1
1965	2
1966	12
1967	1
1968	1
1969	2
1970	2
1971	20
1972	2
1973	20
1974	41
1975	27
1976	120
1977	62

TABLE 8-5

TOP FIVE UTAH STATE LEASEHOLDERS

Leaseholder	Number of Leases	Acreage under Lease	% Total Acreage under Lease
Coastal States Energy	43	82,810	15.2
Utah Resources International, Inc.	46	68,531	12.6
John D. Adams	61	40,221	7.4
Wilene Shumway	56	39,856	7.3
Alumet	25	27,989	5.1
TOTAL	231	259,407	47.6

leases contributed 89,000 tons to the production of Western States Coal Corp.'s Dog Valley mine. United Nuclear Corp. subsidiary, Plateau Mining Co., operates the Wattis mine which stretched onto state land last year.

Two leases owned by U.S. Steel also produced coal in 1977. One contributed to the output of the Horse Canyon mine. High quality coal mined there is burned in the company's Geneva steel mill. A second active lease controlled by U.S. Steel is actually part of Kaiser Steel Corp.'s Sunnyside mine. In 1976 Sunnyside mine reached the boundaries of a lease tract owned by Kaiser. Kaiser then worked out an agreement with U.S. Steel, enabling it to continue its operations onto U.S. Steel's leased land. U.S. Steel retains ownership of the lease and actually pays the production royalties. In return, Kaiser pays U.S. Steel for the rights to the coal.

Active leases on Utah state land produced a total of 320,028 tons in 1977, generating $44,909 in royalties or 14¢ per ton of coal mined. Accepting such low royalty payments was never the intention of the state leasing program. Indeed, it, like others in the West, was established to maximize the revenues accruing to specific institutions Congress considered critical to the state's welfare. Most acreage was awarded to the public school system. The remainder was distributed among an assortment of hospitals, universities and other public facilities (see Table 8-6).

Each year the Division of State Lands collects, records and distributes income generated by activity on state-owned land. Rent payments are turned over to the institutions to be incorporated into their operating budgets for the coming year. In Fiscal 1977, the state collected $29,784 as rent for its leased coal lands. Royalty and advanced royalty payments—which totalled $91,209 in 1977—are deposited in the state's Land Trust Fund and invested. Investment earnings are allocated to the institutions, but the principal is to remain as a permanent asset. The Fund principal has grown into the second largest in the West with a balance of $118 million.

TABLE 8-6

BENEFICIARIES OF UTAH STATE LANDS

Institution	Acreage	Fund Balance
School	3,489,461	$111,484,682.27
Utah State University	25,450	1,107,549.95
Deaf & Dumb	5,931	413,586.82
Utah State Hospital	3,378	550,042.52
Institution for the Blind	19,178	782,915.46
Miners Hospital	2,953	297,803.64
Public Buildings	3,358	133,255.54
Normal School	7,035	539,557.57
State Industrial School	1,590	326,653.14
Reservoirs	52,936	829,975.03
School of Mines	6,163	449,628.63
University	19,085	1,074,610.43
TOTAL	3,636,518	$117,990,261.00

Coal development has contributed only minimally to the Land Trust Fund's growth. In the absence of a rigorous competitive leasing program and with the lack of mining activity on state land, the bulk of the coal revenues accruing to the state are in the form of rent and advanced royalty payments. These sources contribute less than 1% of the Division of State Land's annual receipts from all sources. With barely a shrug, Prince rationalizes the meager support coal leasing provides to the institutions. "There really isn't too much difference. If it didn't come from us, it would come from taxes." Perhaps this is true, but certainly the Division of State Lands has lost its opportunity to garner revenue from a planned coal development program by haphazardly leasing the state's coal reserves at giveaway prices. The benefits of future competitive interest in Utah state coal will be reaped in the private sphere by the various leaseholders.

DEVELOPMENT

Utah coal fields have featured prominently in the history of western coal mining. Since recorded mining began in 1870, coal mines in the state have produced 359 million tons. In 1977, however, its 8.8 million tons represented only 7.5% of all western production for the year, suggesting that Utah is no longer as vital to the western coal scene as it once was.

Recent corporate developments, however, indicate that the coal boom forecast for the West will not bypass Utah. There are currently 17 active coal mines in the state, but a 1976 Bureau of Mines study indicated that 29 new mines or major mine expansion projects are on corporate drawing boards. Utah Power

& Light Co., for example, has announced plans to construct two additional 400,000-kilowatt coal-fired generating units in its Emery power plant, a project that will cost over $620 million and draw upon coal the company will be mining from its Wilberg mine. Pacific Gas & Electric Co. hopes to use six million tons of Utah coal a year to feed a 1,600-megawatt plant it plans to build in central California. Ada Resources, Inc. has formed a subsidiary to supply Utah coal to Texas Gulf Coast industrial plants replacing natural gas with coal to generate steam for various industrial processes. Intercoast Coal Co. plans two mines for Utah—one to open this year—to supply industrial markets in Texas, Arizona and Nevada.

But the biggest news in Utah's coal business this past decade has been the battle over the construction of a mammoth 3,000-Mw coal-burning power plant to supply electricity to a consortium of utilities in Arizona and southern California. The Kaiparowits Project, on the Kaiparowits Plateau in southern Utah, was to have been the biggest coal-fired power plant in the world. Conceived as a 6,000-Mw plant costing $750 million, the plant was jointly proposed in 1963 by Southern California Edison, San Diego Gas & Electric, and Arizona Public Service. The plant was to be fueled by four underground mines with a combined output of 12 million tons a year. A high power 500-Kilovolt transmission system with a supporting communication network spanning 1,460 miles and four states would have been constructed to assure delivery of the power to its markets. A new highway was to be built. A new town to house construction workers and other project personnel was also included in the plan.

By the mid-1970s, the Kaiparowits Project provoked such hot controversy that it grew into one of the largest environmental battles in the West. Stung by the ultimate defeat of their efforts to prevent construction of the Four Corners, Mojave and Navajo generating stations—all of which now supply power to the Los Angeles region—environmentalists went all out to stop Kaiparowits. They claimed that the plant would emit 52 tons of sulfur dioxide, 14 tons of soot and dust, and 250 tons of nitrogen oxides daily. They argued that the project would pollute the pristine air surrounding wilderness and national park areas in the state. On this point they received important support from the Department of Interior's Environmental Impact Statement on the project, which concluded that $600 million of pollution abatement equipment still would not counteract "reduced visibility and sky discolorations due to stack emissions," or the "high adverse impact on the quality of recreation in the Glen Canyon and Bryce Canyon regions." The utilities backing Kaiparowits replied that the power they would provide was much needed in the mushrooming Los Angeles and southern Arizona metropolitan areas. An assortment of businesses, labor unions and state government officials asserted that the construction and operation of the plant would give a much needed lift to Utah's depressed economy.

During the 10 years that the battle over the project raged, costs quintupled to $3.5 billion for a plant half the original size. In April 1976, faced with growing costs, legislative and regulatory delays, environmental lawsuits and the anticipation of more of the same, the two California utilities withdrew from the project, still needing 220 permits and authorizations from 42 federal, state and local agencies.

Plans for western power megaplants never seem to die, however. Rather, they are often reborn, renamed and moved up the road a couple of plateaus. Seventy miles northeast of the Kaiparowits site, but still near several national parks, monuments and recreational areas lies Salt Wash, Utah, the first proposed site for the Intermountain Power Project (IPP). In many ways IPP is the offspring of Kaiparowits. It is a $3 billion plus project proposed by a consortium of utilities, mainly from southern California—though not the same as those behind Kaiparowits. It would bring jobs and revenue to Utah while supplanting 35 million barrels of foreign oil burned in the Southwest annually. It would be one of the biggest coal-burning plants in the world. It would require $600 million in pollution control equipment. Despite the pollution controls, it too would deleteriously affect the environment near several national parks. Its sponsors still face complex and lengthy legal and regulatory procedures.

IPP faces an added hurdle, however. In order to operate legally it will have to obtain an exemption from provisions in the Clean Air Act Amendments of 1977 that prohibit deterioration of air quality in regions surrounding national parks. Secretary of Interior Cecil Andrus has publicly stated his opposition to such an exemption.

Andrus agrees, however, that IPP would supply burgeoning metropolitan centers with needed energy, and he has pledged to support the project if an environmentally-acceptable site is selected. He created a special task force to explore the feasibility of two alternative sites. The first, near Hanksville, also close to Canyonlands National Park, was widely criticized by environmentalists. The second site, near Lynndyl in the center of the state, now appears a more likely choice.

Once a thriving cattle town, Lynndyl's population has been shrinking since the 1940s to less than 100 today. IPP's presence would reverse that trend by creating an estimated 550 permanent jobs and generating $32 million in county taxes—more than 16 times the present yield. Many local residents eagerly await the town's recovery. Merlin Christianson, President of the town's Mormon diocese, relates that "we have prophesies from high officials of the Church that this place will be the hub of wealth."

Some local farmers and ranchers, reluctant to give up their land to light up Los Angeles, view IPP differently. They assert that IPP would return 15,000 newly cultivated acres of farmland to its original sagebrush terrain. Their already dry lands would be drained of 45,000 acre feet of water each year. These traditionalists are making it difficult for the utilities backing the project. They will only sell their water rights to IPP at a rate so high that it will eventually cost the project as much as $225 million to obtain rights to all the water needed. IPP representatives complain that "the farmers are trying to get rich off their water."

Still, the utilities seem determined to push ahead with the Lynndyl site. They have already filed an application with the Department of Interior to build the project there. Secretary Andrus has announced his tacit approval of the plan. An Environmental Impact Statement is now being prepared. Its completion should mark the beginning of a new stage in the effort to bring mammoth coal development projects to Utah.

STATE POSTURE

Utah's state government has long championed industrial development within its borders while lambasting federal interference in the private sector. Recently, it has taken limited strides to regulate coal development. Yet CEP found that Utah still lags behind other western states in its capacity to meet and control the inevitable problems associated with rapid coal development.

Utah is the only major coal-producing state in the West which does not levy a coal severance tax. A limited ad valorem tax is the only levy against coal developers in the state. Each county assesses different ad valorem rates against the value of buildings and equipment on the property. Until 1975, Utah coal mining operations located on federal land—as most of them are—were exempt even from this tax. Federal leaseholders remain exempt from taxation on the value of their coal.

The Utah Legislature is currently considering a bill authorizing a coal severance tax. Robert Cooper, Director of the Mineral Properties Division of the Tax Commission, told CEP chances are slim it will pass committee. According to Cooper, state legislators fear that the coal industry "won't stand for a new tax" so soon after they were forced to begin paying ad valorem taxes on their federal coal lands. Cooper himself suspects that the new tax would make Utah coal less competitive with other western coals, though in the same breath he acknowledges that the state has the most lenient coal tax structure in the West. Cooper's own division does not recommend the severance tax. "We're overworked already," he complained. "We have enough problems. We certainly don't need any more." At the root of this departmental resistance is Director Cooper's own personal philosophy. "I don't really believe in taxes," he confessed.

Partly because it lacks adequate funding and partly because it has yet to part with the state's tradition of extending the welcome mat to industry, Utah's recent efforts to regulate coal development have been ineffective. The state legislature passed strip mining control legislation that falls short of federal guidelines. Its enforcement has been tacked on to the responsibilities of the Division of Oil, Gas, and Mining. When the battle over IPP raged most fierce, the legislature created the Energy Conservation and Development Council to investigate alternative sites, but neglected to empower it with veto power over industry selections.

Utah Governor Matheson promises that his state will "affirmatively develop its own resource policies and not simply react to federal policy." Yet CEP believes that Utah's efforts to regulate developers ultimately seek not so much to control them as to prevent that control from being exercised by others with more exacting standards.

9. WYOMING

INTRODUCTION

"Here in Wyoming, we have been blessed by the Lord with coal," Oscar Swan, Deputy Commissioner of Public Lands in Cheyenne, boasted to CEP. To be sure, Wyoming is endowed with low sulfur, low ash, strippable, sub-bituminous coal in virtually unparalleled quantities, with a strategic proximity to a reliable rail network. This confluence of factors now makes Wyoming the focal point of current western coal development. More than any other western state, Wyoming is experiencing a bona fide coal "boom." Whether or not it is a blessing, however, remains to be seen.

Buried under 25.6 million acres—41% of Wyoming's land area—lie an estimated 936 billion tons of coal, roughly one-fourth of the nation's total coal reserve. The federal government owns the bulk of the coal here, controlling 11.8 million acres or 46% of the coal land. Control of 10.6 million acres is divided among mining companies, railroads, ranchers, utilities and others in the private sector. About 1.9 million acres of coal land lie on the Wind River Indian Reservation, though none has been leased for development. The state of Wyoming owns the remaining 1.3 million acres of coal land.

Although just 6% of the state's coal is considered to be economically recoverable with today's mining technology, even this small percentage translates into a whopping 53.3 billion tons of recoverable reserves, 78 times the nation's total production in 1977. This coal is found roughly equally in strippable and underground seams. Total recoverable reserves in Wyoming are the second largest in the West, topped only by neighboring Montana.

Nearly all of Wyoming's coal, like most coal in the Northern Plains, is sub-bituminous in grade. Together, Montana and Wyoming account for more than 95% of the sub-bituminous coal reserves in the West and 88% of those in the entire country. Fully 99% of Wyoming's sub-bituminous coal contains less than 1% sulfur; more than 50% contains less than 0.7%.

Wyoming also contains about 4.5 billion tons of recoverable, higher quality bituminous coal, but it is sparsely distributed under coal fields in the south-central and southwestern portions of the state. Although bituminous coal is only a tiny fraction of its total coal reserves, Wyoming has more than any other western state except Colorado.

Northeastern Wyoming's Powder River Basin is by far the largest of six distinct coal regions in the state. It alone contains more coal than all but two *countries* in the world. Five million acres there in Converse and Campbell Counties, the majority federally controlled, cover 54% of Wyoming's total coal reserves. The Hanna coal field in southcentral Wyoming contains underground and surface, low sulfur, sub-bituminous coal as well as scattered bituminous deposits. The state's highest grade of bituminous coal underlies the Ham's Fork and Rock Springs coal fields in southwestern Wyoming—the earliest sites of coal development in the state. Reserves have yet to be fully determined for the state's other two coal regions, central Wyoming's Wind River field and northcentral Wyoming's Big Horn Basin field where current production is limited to local needs and no development is imminent.

Coal mining in Wyoming's fields began in the 1850s. Construction of Wyoming's first railroads shortly thereafter encouraged early miners throughout the state; underground operations began on the Rock Springs, Ham's Fork and Powder River coal fields. "From day one," according to Gary Glass, Staff Coal Geologist of the Geological Survey of Wyoming, "the railroads were our primary consumers." As Union Pacific rail traffic gradually accelerated, so too did Wyoming coal production. Production jumped during the World War II years due to the increased pace of railroad traffic, until 10 million tons of coal were produced in 1945—a record that was to last until 1972. Coal production gradually declined in the years immediately following the War, as railroads switched from coal to diesel oil to propel their steam locomotives. A record low in this century for the Wyoming coal industry was established in 1958, when the only consumers were a few small state power and industrial plants.

Revival of the Wyoming coal industry to its current position of pre-eminence occurred in three stages. First, a gradual growth in coal production resulted from the application of newly available strip mining technologies throughout the 1960s. State coal output grew from two million tons annually in 1960 to over four million tons per year in 1969. The second phase accompanied the completion of several power plants within the Northern Plains in the late 1960s and early 1970s. Production in 1970, for example, was 50% higher than in 1969 because of coal demand created by the newly-completed Dave Johnston power plant in Converse County. The third phase ensued when power plants throughout the midwestern and southcentral states began buying Wyoming coal. This added a huge export market to the state's coal economy. In 1969, 75% of Wyoming's coal was used in-state. By 1976, only about one-third of the coal produced was bought by in-state customers. Total coal production during this period skyrocketed from 4.6 to 31.1 million tons.

Sixteen active coal mines brought Wyoming production up again to 45 million tons in 1977, a 41% increase over statewide output the previous year. This level of production exceeded by 18 million tons the 1977 coal production of Montana, the western state with the second largest output. Its now booming coal fields have thrust Wyoming into the forefront of all western coal-producing states. The 596.7 million tons of coal that have emerged from Wyoming's coal mines surpass the historical coal production of each western state except Colorado, the traditional center of the western industry for over a century. By the end of 1978, Wyoming's historical coal production will exceed even Colorado's.

Continued strip mining of Wyoming's easily accessible, low sulfur, sub-bituminous reserves for export will be at a rate swift enough to solidify Wyoming's preeminence among western coal-producing states. The Department of Energy forecasts that Wyoming coal fields will produce as much as 211.5 million tons per year by 1985. This expectation represents a quintupling of the record 1977 output. By 1985, Wyoming could account for over 42% of all western coal production. The nationwide demand for Wyoming coal that helped revive the state's coal industry will by then swell exports to over 80% of state output. The Burlington Northern and Union Pacific Railroads, whose switch from coal to diesel oil once threatened the demise of the Wyoming coal industry, will subserve coal development expansion here by supplying Wyoming coal to power plants in 16 states.

FEDERAL LEASING

Of the more than 25.6 million acres of coal-bearing land in Wyoming, the federal government owns 11.8 million acres, or 46%. Only Montana contains more total federal coal acreage than Wyoming.

Ninety-six federal leases spanning 214,821 acres have been issued for development of Wyoming's federal coal resources. Over 98% of the federal coal acreage in Wyoming remains unleased. Leases in Wyoming are located in three general regions. The 132,000 acres of leased land in northeast Wyoming's Powder River Basin constitute almost 62% of all federal coal lands under lease in the state. Southcentral Wyoming's coal fields in eastern Sweetwater and Carbon Counties contain over 26,000 acres of leased federal coal lands involving 28 leases. Twenty-one leases covering over 55,000 acres of federal land have been issued for development on the coal fields of south-western Wyoming's Uintah, Lincoln and Sweetwater Counties.

Despite the federal moratorium on leasing, activity on federal lands in Wyoming during the past five years has been among the most extensive in the West. One new preference right lease has been issued to Rosebud Coal Sales Co. in February 1976 for development on a huge 14,900-acre tract in Sweetwater County. In addition, four new leases have been "segregated" from already existing leases. Total acreage under lease has consequently increased by 7% from 199,898 acres in 1973, representing the most significant federal lease acreage growth in the West.

Furthermore, 37 federal coal leases in Wyoming, representing 39% of all federal coal leases here, have been assigned to new owners since 1973. In the West, only New Mexico has experienced a greater percentage of lease assignments.

Wyoming's leased federal lands contain nine billion tons or 56% of all federal coal reserves under lease in the West. In addition, 8.4 billion tons of coal reserves underlie Wyoming's federal acreage currently under preference right lease application. Such extensive reserves—more than 80% of which are strippable—represent 63% of all federal reserves committed to lease in the West.

The revitalized pace of coal mining in Wyoming has caused dramatic increases in total federal coal production there. In 1973 only nine Wyoming federal leases were active, yielding approximately seven million tons of coal annually. By 1977, twice as many federal leases were active, generating four times as much coal as in 1973, accounting for fully 55% of all 1977 federal coal production in the West. The 28.7 million tons of federal coal mined in Wyoming in 1977 is nearly three times as much as the federal output in Montana, the second largest producing state. Furthermore, two of the ten largest coal mines in the country in 1977, the Belle Ayre and Jim Bridger mines, are located on federal land in Wyoming.

The Department of Interior projects that Wyoming's federal coal output will increase to just under 100 million tons per year by 1990—more than twice as much as the projected federal output of any other western state, and 35% of the West's total Department of Interior production forecast for that year. Other more aggressive estimates indicate that in the Powder River Basin alone, nine currently licensed mines situated on 45,663 federal acres containing almost four billion tons of reserves are slated to produce between 88 and 121 million tons per year by 1985. The Bureau of Land Management (BLM) will soon be considering 19,000 acres of federal coal land in the Powder River Basin for new leasing, and another 10,500 acres for supplementing existing lease holdings—some of which are slated to produce as much as 25 million tons per year by 1985.

Dramatic growth in federal coal output in Wyoming seems inevitable. Still, surface-subsurface conflicts exist there, and will have to be resolved before mining plans on federal land can be approved. Of the more than 214,000 acres of federal land now under lease, roughly 60% are privately owned at the surface, so subsurface-surface disputes are sure to occur. They will not rage as fiercely, however, as in states like Montana and North Dakota, where virtually all federal mineral lands are privately owned at the surface.

Such conflicts will exist in Wyoming as a reflection of the general surface-subsurface entanglements that beset the West at large, where the federal government transferred surface ownership to private parties but retained the minerals underneath for itself. These disputes will do little to curtail the unparalleled growth in coal output forecast for Wyoming's federal lands, however. American Electric Power Co. (AEP) has filed suit to halt Kerr-McGee's coal mining on a 350-acre tract controlled by AEP at the surface. The utility holding company has described Wyoming's state law permitting the mineral rights owner to mine in defiance of the surface owner's disapproval as an unlawful deprivation of property.

STATE LEASING

Sometime in the mid-1950s, at least 10 years before the emergence of a competitive interest in Wyoming coal, two citizen advisory committees were established within the state's executive branch. One was sent east, the other west, to search out markets for Wyoming's sub-bituminous coal. As Deputy Commissioner of Public Lands Oscar Swan remembers: "Neither group could sell a bucket."

Such experiences heavily influenced Wyoming's state coal leasing policy. Convinced that no enthusiastic pursuit of Wyoming coal would ever occur, the state began giving away its coal land. No competitive lease sale was ever held; no cash bonus payment ever received. Regardless of size, each lease was issued for only $15. Rental, royalty and advanced royalty rates were established that were among the lowest in the West. No measures were taken to curb speculation or promote development on state resource-bearing acreage. Fifteen years after the two groups of Wyoming coal ambassadors returned empty-handed, the state's Deputy Commissioner of Public Lands told CEP: "In the sense of having any plan for coal development, we don't." CEP agrees; the state of Wyoming operates the most grossly inadequate, and maladministered leasing program in the West.

The state of Wyoming owns roughly 5% of all the coal land lying within its borders. This land—1.3 million acres—is part of the 4.2 million acres deeded to the state by the federal government when Wyoming entered the Union in 1890. The state was awarded two, noncontiguous sections (16 and 36) each one square mile, in every 36 square mile township. Occasionally, sections 16 and 36 were under private or Indian control prior to Wyoming's admission to the Union, so the federal government permitted the new state government to choose other sections in exchange. Through these exchanges, the state was able to acquire some tracts larger than 640 acres. But for the most part, state-owned land in Wyoming occurs in a checkerboard pattern of small plots intermingled with land owned by the federal government, Indian tribes or private individuals.

The 1.3 million acres of state coal land are estimated to contain between 13 and 16 billion tons of recoverable coal reserves, making Wyoming the largest coal owner by far among the six western states examined by CEP. Roughly 90% of the state's coal is sub-bituminous in grade. Nearly half of the state's coal resources lie in the Powder River Basin; another quarter are in the Green

TABLE 9-1

WYOMING STATE LEASE ISSUANCE

Years	Number of Leases Issued
1955-1959	7
1960-1964	23
1965	378
1966	153
1967	14
1968	124
1969	18
1970	12
1971	102
1972	128
1973	205
1974	70
1975	39
1976	180
1977	115

River coal region in the southwestern portion of the state.

Since 1957, the Board of Land Commissioners within the Department of Public Lands and Farm Loans has administered the coal leasing program on state-owned land in Wyoming. The Board has issued 1,568 leases—more than three times the number of leases issued in Utah, the western state with the second largest state lease catalogue. These 1,568 currently valid leases comprise 1,235,229 acres, more than twice the acreage under lease in Utah and 57% of all leased state acreage in the West. Virtually all of Wyoming's coal-bearing land already has been transferred to the private sector. The state then no longer really operates a leasing program but a "lease maintenance" program.

Lease issuance in Wyoming has centered around three broad boom periods in contrast to most western states which only recently have experienced dramatic resurgences of leasing activity (see Table 9-1). The first leasing flurry occurred between 1964 and 1966 when Carter Oil Co. and Mapco, the two largest holders of state leases in Wyoming and in the West at large, bought up rights to more than 250 leases. Between 1971 and 1973, in the early days of the federal coal leasing moratorium, energy companies and speculators turned their attention to Wyoming's state coal leasing program and created a second leasing boom there. Since 1976, prospective lessees have again looked to state coal reserves in Wyoming and accounted for a third leasing boom. Approximately 39% of all Wyoming state coal leases have been issued in the past five years.

TABLE 9-2

REVENUE FROM WYOMING STATE LEASE SALES

Method of Leasing	Number of Leases	Total Revenue Collected
Open Land	1,568	$23,520
Segregation	0	0
Competitive Bidding	0	0
TOTAL	1,568	$23,520

Throughout its history, the Wyoming Board of Land Commissioners has offered prospective lessees the most generous lease terms in the West. Wyoming is the only state examined by CEP that has never issued a single lease competitively. All of its leases have been issued upon payment of a mere $15 filing fee. The state has collected a total of $23,520 or just 2¢ per acre for the issuance of all its coal leases (see Table 9-2). In a classic understatement, Deputy Commissioner Swan told CEP that "the cash consideration in leasing is nominal."

A.E. King, Commissioner of State Land in Wyoming, described this "open land" leasing procedure to CEP in an interview in December 1977: "We are the only state where land is open to the first qualified applicant." A qualified applicant is any person at least 21 years old who is a citizen or has declared an intention to become a citizen of the United States, or any firm or corporation which has complied with the laws of the state. In order to obtain a lease, a qualified applicant must appear at the State Land Office in Cheyenne, search the land plot maps until a section of unleased land is discovered, fill out a simple application form, and pay a filing fee of $15. If the applicant can prove that the land has not yet been leased, if the application form is correctly completed, if the $15 filing fee is paid, and if the first year's rental payments are rendered in advance, the Commissioner must issue the coal lease.

When land covered by a relinquished, cancelled or terminated lease becomes available for releasing, a different issuance procedure is employed. The state issues a public notice of a "simultaneous filing" period announcing the availability of the land for releasing. Prospective lessees may then file an application to obtain the lease tract. Within 10 to 30 days after the notification of the availability of the land, a lottery drawing is held. Cards, each containing the name of one applicant, are placed in a metal drum. The drum is revolved and a card is drawn. The applicant whose card is drawn receives the lease for a $15 filing fee.

Those prospective lessees who do not want to deal with the open land and lottery leasing systems, are able to purchase leases without evan talking to a state official. Wyoming, like all other western states, permits leaseholders to sell their lease tracts on the open market to the highest bidder. Fully 27% of all outstanding coal leases have devolved into the hands of their current owners by assignment. Wyoming receives no share of the cash bonuses which accompany assignments. The only control the state exercises over the assignment process is its limiting of overriding royalty provisions to 5% of the gross value of the coal.

Wyoming has been as lackluster in generating revenues from the maintenance of leases as from lease issuance. The state employs the lowest rental structure in the West. Leases issued prior to 1974 contained a rental requirement of 25¢ an acre per year for the first five years and 50¢ an acre for years six through 10. On January 11, 1974, this rental rate was doubled to 50¢ and $1 per acre per year. Even this new rate is among the lowest in the West. In Fiscal 1977 the state received $336,000 in rent payments—an average of 27¢ per acre, by far the lowest average rent collected by any western state government.

Lessees who report the existence of commercial quantities of coal on their lease tracts are obliged to pay advanced royalties of $1 per acre per year. These advanced royalties are credited against rental payments, so the lessee's total payment seldom changes; instead of paying $1 per acre rent, the lessee pays the same amount in advanced royalties. But because advanced royalties, unlike rent, accumulate as credit against future royalties, most leaseholders declare the existence of commercial quantities of coal and begin paying advanced royalties as soon as rental payments reach $1 per acre—usually in the fifth year of the lease. The leaseholder is under no burden to actually prove the discovery of coal, nor does the state attempt to verify it. Advanced royalties collected in 1977 exceeded $900,000.

On state lease tracts which actually produce coal, lessees are required to pay royalties on a per ton basis. Here again, Wyoming's assessed rates and collected total payments are among the lowest in the West. Most older leases charge a maximum of 5¢ per ton, the lowest royalty rate encountered by CEP. Leases issued in the last few years generally charge 8% of the value of the coal—still far below the federally-mandated minimum—or 25¢ per ton. Furthermore, until 1977, only one Wyoming state lease had ever produced any coal at all, thereby leaving the state with only a limited source of royalty income.

Three active leases produced 740,000 tons of state coal in 1977 and generated royalties of $124,000—an average rate of 17¢ per ton. Roughly 400,000 tons of the 2.6 million tons produced at Rosebud Coal Sales Co.'s Hanna mine emerged from a state tract—one of the oldest state leases in Wyoming and the only currently valid lease which produced any coal prior to 1977. Since 1965, this lease has yielded more than two million tons of state coal. Kemmerer Coal Co. owns a state lease in southwestern Wyoming's Lincoln County which began producing in early 1977 and promptly achieved an annual output of 240,000 tons. Ark Land Co.'s state lease contributed 90,000 tons to its extensive operations in Carbon County.

That less than 0.2% of all valid state coal leases in Wyoming are now in production suggests the Department of Public Land's failure to establish sufficient incentive to encourage development. The state has never exercised its authority to make renewal of lease terms contingent on the lessee's capacity to prove the imminence of "diligent production" on the lease. The state leasing legislation specifies that after the expiration of an inactive lease's terms—state leases in Wyoming have a 10-year duration—the state may conduct competitive auctions and force the lessee to match the highest competitive offer for the lease tract before granting renewal rights. Wyoming, however, has never offered a lease for competitive bidding as part of the renewal process. Instead, each expired lease has been automatically renewed at the request of the lessee.

Ownership of Wyoming's 1,568 leases is scattered among 298 leaseholders—each controlling an average of 5.2 leases. A small assortment of giant energy developers and land speculators controls most of Wyoming's coal reserves, however. The top five leaseholders control about 43% of all coal land under lease, an average level of concentration among the western states (see Table 9-3). Exxon subsidiary Carter Oil Co., the largest leaseholder in the West, is also the largest state leaseholder in Wyoming, with 155 leases covering 153,904 acres. Carter Oil purchased all but 10 of these coal leases on one day in April 1965 for a total payment of only $2,175. On that same day, Mapco—the largest state leaseholder in the West and the second largest in Wyoming—obtained 109 of its 149 coal leases. Ark Land Co., the state's third largest leaseholder, has acquired all 90 of its coal leases by assignment. So, too, the fifth largest leaseholder, P.C. Development, Inc. obtained its 97 leases by assignment.

R.A. Haynesworth, a Cheyenne land broker, controls all or part of 258 leases. He has outright ownership of only seven of these, holds 239 in joint ownership with one other party, 10 in tri-party ownership, and two in a team of four. CEP has computed the acreage controlled by Mr. Haynesworth to total 82,059, making him the state's fourth largest leaseholder. He has a strong influence on the disposition of an additional 81,484 acres—the acreage controlled by his various co-lessees.

These co-lessees are scattered across the country from Boston suburbs to Hawaii. This geographic distance makes it difficult, CEP suspects, for

TABLE 9-3

TOP FIVE WYOMING STATE LEASEHOLDERS

Leaseholder	Leases	Acreage under Lease	% Total Acreage under Lease
Carter Oil Co.	155	153,904	12.5
Mapco	149	139,878	11.3
Ark Land Co.	90	83,615	6.8
R.A. Haynesworth	258	82,059	6.6
P.C. Development, Inc.	97	69,606	5.6
TOTAL	749	529,062	42.8

Haynesworth to keep in touch with his lease partners. There is, in fact, some evidence that he is less than rigorous in keeping his co-lessees abreast of their involvement. A State Department of Public Lands employee told CEP that several of Haynesworth's partners wrote or appeared at the land office requesting verification of their rights to coal leases. Apparently, they had been contacted by Haynesworth, who convinced them to enter the joint venture. After paying him their share of first year's rent, they never heard from Haynesworth again.

Haynesworth is a less elusive figure in Wyoming, it seems. He often appears at the state land office on a daily basis, accompanied by W.M. Wilson, another local land broker who is the state's sixth largest leaseholder in his own right, controlling all or part of 105 state leases. Haynesworth is newer to the state leasing scene in Wyoming than Wilson. His leases have all been obtained since 1971. He purchased 107 of them from the state on one day in August 1976, and another 54 on April 1, 1977. His other leases were also all acquired directly from the state.

Low lease issuance and maintenance fees and antiquated land management policies have been Wyoming's feeble responses to the Congressional directive to maximize revenues generated from activity on state land. Upon admitting Wyoming into the Union, Congress selected 16 institutions as the sole beneficiaries of such revenues (see Table 9-4).

The Land Consolidation Act of 1975 slightly altered the dispersal of funds to these institutions. Those with only small state acreages were grouped together. Rental payments for land assigned to these institutions were placed in the state's General Fund rather than distributed among the institutions. Guy Sturlin of the Department of Public Lands' Accounting Office told CEP this reform was to simplify accounting practices and not to shortchange the institutions. These small institutions now draw their support from the general fund. Larger institutions, such as the common schools, the university, and the agricultural college, continue to receive rental payments directly. Their share in 1977 totalled $336,000.

Royalty and advanced royalty payments go neither to the institutions nor to the state's General Fund but into the state Permanent Fund, where they commingle with similar payments collected from oil, natural gas and uranium activities on state land. The principal of this fund is invested, and the income is then turned over to the institutions. Wyoming's Permanent Fund has a balance of $224 million, the second largest in the West. Contributions from coal development in 1977 totalled $1,025,000—90% of which came from advanced royalties. Coal activity generated 5% of the total revenue accruing to the Fund in 1977.

Benefits to the state of Wyoming from the leasing of its coal land accrue simply because of the sheer magnitude of coal resources owned and leased by the state, not because of the sagacity of the state leasing agencies. Management policies to guide the state leasing program are virtually non-existent. No attempt is made to control the rate or location of development.

TABLE 9-4

BENEFICIARIES OF WYOMING STATE LANDS

Institution	Acreage
Common School	3,543,599.48
University	47,765.19
Agricultural College	88,459.47
State Charitable, Educational, Penal, and Reform Institutions (Omnibus)	200,048.29
Fish Hatchery	6,095.78
Deaf, Dumb and Blind Asylum	29,332.39
Poor Farm (Wyoming State Training School)	9,474.78
Penetentiary	30,125.43
Public Buildings at Capitol	87,586.18
Erection of Public Buildings at Capitol	31,436.79
Penal Reform or Educational Institution	27,959.31
Soldiers' and Sailors' Home	31,100.43
State Law Library	14,581.04
State Library	14,692.52
Insane Asylum	29,871.52
Miners' Hospital	30,006.25
TOTAL	4,222,134.85

There is little interest in interfering with leaseholders at any level. Oscar Swan defended Wyoming's coal leasing program, telling CEP that: "No one has ever been able to show us that our system is not as good as theirs." CEP believes our analysis does that, at last.

DEVELOPMENT

More than half of Wyoming's coal production in 1977 emerged from Powder River Basin mines in Campbell, Converse and Sheridan Counties—most on federal lands. By far the largest was AMAX's Belle Ayre mine—the country's largest coal mine in 1977—situated on a federal tract in Campbell County. Production there exceeded 13 million tons last year, just one-fourth of the coal production for the entire state. Pacific Power & Light Company's Dave Johnston mine produced over three million tons in 1977, and Big Horn Coal Co. expanded production to more than two million tons. Each in its first year of operation, Sunoco Energy Development Co.'s Cordero mine and Carter's Rawhide mine generated more than two million and one million tons respectively in 1977, all on leased federal land.

Roughly half of the currently planned new mines and expansions of existing mines in Wyoming will occur in the Powder River Basin. The Powder River Basin Resources Council, a coalition of ranchers and environmentalists, suspects that output there could reach as high as 125 to 175 million tons per year by 1985 from federal lands alone, not even including pending lease expansions. The Belle Ayre mine should increase production from 13 million to 20 million tons annually. The Cordero mine will expand production to 12 million tons per year by 1981 and eventually approach a 14 million ton per year capacity. The Rawhide mine is slated for 1980 production of 8.5 million tons, two-thirds of its capacity. A smaller Wyodak Resources Development Co. strip mine has been cleared for a five-fold increase in production from 800,000 tons last year to four million tons per year in the 1980s.

Permits have been issued for the operation of mammoth new Power River Basin strip mines to complement these expansions. AMAX's Eagle Butte mine will start production this year and will quickly reach its 20 million ton per year capacity. Kerr-McGee will bring into operation this year its Jacobs Ranch strip mine slated to produce 14 million tons annually by 1983. Atlantic Richfield's Black Thunder mine—scheduled to yield 10 million tons per year by 1980—is reportedly on the verge of production. Exxon's Carter Oil too will soon begin operations on its Coballo property south of Gillette, hoping to produce close to five million tons per year by 1980 and 12 million tons annually at capacity.

Should gasification plants ever be constructed and operated there, the Powder River Basin region will attract even more coal mining. Peabody Coal Co. has leased property in southern Campbell County, from which it hopes to mine 11 million tons per year at capacity to feed a gasification plant in Douglass, Wyoming. Similarly, Texaco has announced plans for a strip mine in Johnson County to supply coal for an adjacent gasification plant it hopes to construct.

Boom forecasts for the Wyoming coal industry have stimulated similar development on coal fields throughout the state. Traditionally a backbone of the state's coal industry, Carbon County's coal fields in southcentral Wyoming, accounted for more than one-quarter of Wyoming's 1977 coal output. Five mines here, each yielding well over one million tons, produced more than 12 million tons last year. Carbon County's large mines included Medicine Bow Coal Co.'s Medicine Bow mine which stripped 2.8 million tons, Arch Mineral Corp.'s Seminoe 1 and 2 mines which produced 2.3 and 2.7 million tons respectively, and Rosebud Coal Sales Co.'s Hanna mine whose 1977 coal output totaled 2.6 million tons. Even the "smallest" Carbon County mine— Energy Development Co.'s Vanguard mine—produced almost 1.5 million tons of coal last year.

Rocky Mountain Energy Co., a Union Pacific Corp. subsidiary, will figure prominently in the expanded output forecast for Carbon County's coal fields. From its underground Hanna mine, it expects soon to produce 2.5 million tons per year. In a joint venture with Arch Minerals Corp., Rocky Mountain Energy plans on eventually stripping three million tons per year from its China

Butte mine southwest of Rawlins, Wyoming. Finally, it hopes to strip 2.5 million tons per year from its Red Rim mine in a joint venture with Energy Development Co., a subsidiary of Iowa Public Service Co.

Energy Development Co., in its own right, is scheduled to increase production from the Vanguard mine near Hanna from 1.5 million tons in 1977 to 2.5 million tons in 1980. By then Medicine Bow Coal Co. will expand its strip mine to a capacity of 3.6 million tons per year—enough to keep it the largest mine in southcentral Wyoming.

A third locus of Wyoming coal activity borders Utah in the state's southwestern corner. There in Sweetwater and Lincoln Counties, four mines produced almost 11 million tons of coal last year—just under a quarter of statewide production. This region's coal emerged primarily from Bridger Coal Co.'s Jim Bridger mine and Kemmerer Coal Co.'s Sorenson mine, the second and third largest mines in the state, which together accounted for just under ten million tons of coal last year. Smaller mines operated by FMC Corp. and Stansbury Coal Co., together produced one million tons of coal.

New mines and scheduled expansions are expected to almost triple production in southwestern Wyoming to roughly 30 million tons annually in the early 1980s. The Bridger mine will expand production to more than seven million tons per year by 1980. By that time, Kemmerer Coal Co.'s operations should be generating as much as six million tons annually. Output from southwest Wyoming's two smaller mines should expand to 3.5 million tons annually by the 1980s. Rocky Mountain Energy Co. hopes to open one new strip mine in Lincoln County generating three million tons per year by 1980 and another, larger strip mine in Sweetwater County in partnership with Peter Kiewit Sons Co. Production from this operation is expected to approach 4.2 million tons annually by 1980.

The demand for Wyoming's strippable low sulfur coal among midwestern, southwestern and Pacific northwestern utilities that inspired the coal boom there show no signs of tapering off. Assuming no new Wyoming power plants or gasification plants come on line before 1980, 85 million tons of the 100 million ton output forecast for that year will be marketed out of state.

Most coal developers in northeastern Wyoming's Powder River Basin, the coal-producing region closest to the Midwest, have entered into long-term contracts with utilities to the east. AMAX has contracts to supply coal from its Belle Ayre and Eagle Butte holdings to utilities in southern, midwestern and Ohio River Valley states. Much of Sunoco Energy Development Co.'s Cordero output will go to public power plants in San Antonio, Texas. Kerr-McGee's two planned mines, Jacob's Ranch and East Gillette, will sell coal to Arkansas Power & Light, Central Louisiana Electric Co., Gulf States Utilities, and Public Service of Oklahoma power plants. Carter Oil Co.'s Rawhide mine will feed coal to Indiana, Michigan and Nebraska power plants; in addition, Carter Oil has recently agreed to a 25-year contract with Western Fuels Association, a cooperative of consumer-owned rural electric and municipal utilities, that will distribute the coal to Missouri Basin Power Project's

Laramie River plant and Public Utilities of Kansas City's new Nearman plant. Atlantic Richfield's Black Thunder mine will yield coal that will travel the same eastward path, supplying power plants operated by Nebraska Public Power District, Oklahoma Gas & Electric Co., and Southwestern Public Service Co.

Little Powder River Basin coal will go to Wyoming customers. Wyodak Resources Development will continue to feed its own Wyodak power plant. Similarly, Pacific Power & Light's Dave Johnston mine will supply the nearby Dave Johnston power plant. And of course, Texaco and Peabody hope to mine Powder River Basin coal for local gasification plants still seeking permits.

Southcentral Wyoming's coal mines will also continue to find their primary markets to the east. The largest operation here, Medicine Bow Coal Co.'s strip mine, will still supply Northern Indian Public Service Company power plants among other midwestern customers. Energy Development Co.'s Vanguard and Red Rim coal will be transported to its parent, Iowa Public Service, for power plant use.

More remote from the midwestern markets, southwestern Wyoming developers are often committed to users within the state and in neighboring states. Bridger mine coal is burned in the adjacent Jim Bridger power plant. FMC Corp. produces coal for a nearby trona plant it operates. Kemmerer Coal Co.'s Lincoln County operations have traditionally supplied the Naughton power complex here, but will soon diversify by providing coal to Allied Chemical and Amalgamated Sugar Co. Stansbury Coal Co.'s coal will also be used for industrial purposes—primarily by Ideal Cement Co.

STATE POSTURE

The rapid pace of coal development in Wyoming has occasioned dramatic changes in the state's social, political and economic complexion, and will exert a profound impact on its human environment. To be sure, many of the effects of coal development will be positive. In the next 10 years, $50 billion worth of coal development activity will bring new schools, service industries and more sophisticated health facilities to Wyoming. Already, per capita income has swelled from $2,897 in 1967—$265 below the national average then—to $6,723 in 1977, surpassing by $282 today's national average. In 1976 alone, total personal income soared upward 24% from the prior year, a growth rate unmatched by any other continental state. Wyoming's current unemployment rate of 2.9% is the lowest in the country.

A few companies, like Atlantic Richfield, have sought to lead communities in planning to meet and comfortably absorb the effects of development. ARCO has invested $8 million in building up services in Wright, Wyoming to deal with the impact of its nearby Black Thunder surface mine. In a similar spirit, the Missouri Basin Power Project formed its own Impact Alleviation

Office in connection with its construction of a coal-fired generating plant in Wheatland.

Still, many experts doubt whether coal development's impact, because of its sheer magnitude, can be contained by these limited precautions. By the year 2000, more than 100,000 acres in the East Powder River Basin alone will have been strip mined. Destruction of surface and underground water systems could make reclamation impossible, given the area's aridity. Several Powder River Basin mines scheduled for opening by 1985 propose to intercept important river and stream beds. Diversion of these streams and reconstruction of their beds could disrupt the equilibrium between the alluvium and the surface flow, alter flow patterns and groundwater recharge permanently, and contribute to silt deposition that may reduce availability of water to downstream users. Data suggests that mines operated by Wyodak Resources and Kerr-McGee, both located east of Gillette, may interfere with that city's entire municipal water supply. Coal mining throughout the region could dry up wells and springs, and saddle rural users with high replacement costs. The Wyoming Game and Fish Department anticipates reductions in local deer, antelope and elk caused by the destruction of their habitat by mining, and the replacement of habitat with industrial facilities, rail lines, roads and home sites.

Foremost on many minds is the concern that the sudden arrival of vast numbers of new residents and the construction of new roads, houses and factories will alter the face of Wyoming permanently, and for ill. The Eastern Powder River Basin can expect 35,000 to 45,000 new inhabitants and more than 9,000 new housing units as a direct result of coal development. Uranium and trona development and new railroad construction will bring into the region another 15,000 to 20,000 residents. Already Campbell County's population has almost doubled from 13,000 in 1970. Mining centers Gillette, Douglass, Newcastle and Glenrock will experience population growth averaging 10% annually until 1985, with growth sometimes reaching as high as 15% per year.

These new residents will lodge greater demands on Wyoming's public services that will weaken the existing structures, raise the cost of living, and skew the balance of revenues collected and expended. A few years ago when construction of new housing could not match the rate of coal development-related immigration in Rock Springs, for example, families temporarily resorted to living in tents. Water and sewer hook-ups in Gillette that cost $75 in 1970 now cost $2,000. Campbell County's school system, which only recently consisted of one elementary school, one junior high school, one high school and one school bus, now has 60 school buses, several new schools and an unprecedented $7 million school bond debt. The next six strip mines to open in the Gillette area will create fiscal demands so extensive that the cumulative annual deficit could reach $92 million by 1990—this in a city historically accustomed to a balanced budget.

Departures will also occur from the traditional political dynamics of the region. Resistance to coal development has already inspired the previously unthinkable alliance of the Wyoming Stock Growers Association and state conservationists to oppose the mining companies in the struggle for water reserves. The presence of multinational corporate giants and their high-powered lobbyists pushing billion dollar coal projects before part-time legislators in a state with very weak financial disclosure laws raises serious questions about Wyoming's capacity to control coal development.

Wyoming has long imitated other pro-development western states like Utah, and refused to interfere with industrial activity. The situation is beginning to change, however, as the state gradually recognizes the incredibly high stakes involved in the coal boom. While Wyoming certainly has not transformed itself into a zealous regulator of energy developers, it has taken a few steps in this direction. Early this decade, for example, the Department of Environmental Quality was formed to enforce strip mine reclamation and regulate activities polluting the air and water. The Department has attained a respectable reputation for its efforts to date. Coal impact offices have been established in other agencies around the state, though they have achieved an uneven record of effectiveness.

Wyoming has enacted moderately steep coal severance taxes to help raise the revenues required to fund its coal impact control program. In 1977 the legislature evaluated the severance tax upward from 4% to 8.5% of the value of the coal. Layered on top of this is a coal impact excise tax authorized to gradually rise from 1.2% in 1977 to 2% of the tax base. The mature coal impact tax together with the severance tax will eventually endow Wyoming with a combined severance tax of 10.5%—the second highest tax on the privilege of coal extraction in the West.

In 1977 coal severance and excise taxes totaled 9.7% of the value of the coal and generated $17.7 million—triple the coal severance tax yield of 1976. The Permanent Wyoming Mineral Trust Fund, a perpetual trust with a current balance of roughly $80 million, received $4.5 million for investment purposes. Interest from these investments is placed in the General Fund, along with an outright grant of over $3.6 million generated by the severance tax. The Water Development Account received $2.7 million from severance taxes which it will call on to guarantee loans assumed by the city of Gillette for construction of a water development project. The State Highway Fund received $1.8 million of the severance tax yield for highway-related expenses to be supervised by the Wyoming Highway Commission.

The recently-established Capital Facilities Revenue Account was granted $2.7 million, three-fourths of which was to be used to fund state capital facilities authorized by the legislature. After it issued its first bond in August 1977 for penitentiary renovation and university construction, however, a state court ruled the pledging of mineral revenues to repay such bonds to be unconstitutional. New uses are now being sought for this portion of the Capital Facilities Revenue Account. The Account's remaining revenues are to be spent for public school and community college construction.

The coal impact excise tax generated almost $2.2 million last year which raised the balance of the Coal Tax Revenue Account it feeds to $2.7 million. This Account was established by the legislature to finance public water, sewer, highway, road or street projects in areas which are affected by coal development. The Wyoming Farm Loan Board administers the Account.

Ad valorem property taxes on coal production on all land in the state also provide the state with coal-related revenues. These revenues are returned to the county where the mining occurred, except for 0.3% which goes into the State Education Fund. While each county levies a different ad valorem rate, the statewide average is 6.1% of the gross value of the coal produced in the previous year. Estimated 1977 ad valorem yield was over $10 million.

Perhaps befitting this western state sparsely populated by independent ranchers who cherish their self-sufficiency, no real state posture has emerged in Wyoming toward coal development. Although severance taxes among the most effective in the West have been authorized, these seem to be designed less to discourage development than to increase the state's share in the resource development it considers inevitable. Unlikely alliances have been struck to protect the range and its wildlife, rural citizens, water rights, and a balanced economy. Still, no sweeping programmatic discouragement of large scale coal development has occurred in Wyoming, as it has in neighboring Montana. On the other hand, Wyoming Governor Herschler recently commented: "We aren't out looking for growth. We've got all the growth we can handle." The more vital issue seems to be whether Wyoming can handle the growth it already has.

ABBREVIATIONS

BIA	Bureau of Indian Affairs
BLM	Bureau of Land Management
BPA	Bonneville Power Administration
CDIO	Coal Development Impact Office
CERT	Council on Energy Resource Tribes
CONPASO	a joint venture of Consolidation Coal Co. and El Paso Natural Gas Co.
EIS	Environmental Impact Statement
EPA	Environmental Protection Agency
EMARS	Energy Minerals Activity Resource System
GAO	General Accounting Office
IPP	Intermountain Power Project
NEPA	National Environmental Policy Act
NRDC	National Resources Defense Council
OPEC	Organization of Petroleum Exporting Countries
REAP	Regional Environmental Assessment Program
SEC	Securities and Exchange Commission
SEICO	Socio-Economic Impact Coordinators Office
USGS	United States Geological Survey
WESCO	Western Gasification Co.
WPI	Wholesale Price Index

NA	Not Available

BIBLIOGRAPHY

COAL DEVELOPMENT—GENERAL

Comptroller General of the United States. *Rocky Mountain Energy Resource Development: Status, Potential, and Socioeconomic Issues.* Washington: The General Accounting Office, 1977.

Comptroller General of the United States. *The State of Competition in the Coal Industry.* Washington: The General Accounting Office, 1977.

Comptroller General of the United States. *U.S. Coal Development—Promises, Uncertainties.* Washington: The General Accounting Office, 1977.

Minger, Terrell J. and Oaks, Sherry D., Editors. *Growth Alternatives for the Rocky Mountain West.* Boulder: Westview Press, 1976.

Tyner, Wallace E., Kalter, Robert J., and Wold, John P. *Western Coal: Promise or Problem?* Ithaca: Cornell University Press, 1977.

United States Department of Interior. *Projects to Expand Fuel Sources in the Western States.* Bureau of Mines Information Circular Number 8719. Washington, 1976.

United States Department of Interior. *The Status of the U.S. Coal Industry.* Bureau of Mines Information Circular Number 8707. Washington, 1976.

* * * *

Bureau of Mines, Department of Interior, Washington.

Department of Energy, Washington.

Energy and Minerals Division, General Accounting Office, Washington.

Materials Division, Office of Technology Assessment, Washington.

Office of Energy Resources, Geological Survey, Department of Interior, Reston, Virginia.

* * * *

Environmental Law Institute, Washington.

National Coal Association, Washington.

National Coal Policy Project, Washington.

FEDERAL LEASING

Cannon, James S. *Leased and Lost: A Study of Public and Indian Coal Leasing in the West.* Economic Priorities Report, vol. 5, no. 1. New York: Council on Economic Priorities, 1974.

Comptroller General of the United States. *Role of Federal Coal Resources in Meeting National Energy Goals Needs To Be Determined and the Leasing Process Improved.* Washington: The General Accounting Office, 1976.

Hearings on *Federal Coal Leasing* before the U.S. House of Representatives Subcommittee on Mines and Mining of the Committee on Interior and Insular Affairs, March 14, 1975. Washington: U.S. Government Printing Office, 1975.

Hearings on *Leasing and Western Development of Coal* before the U.S. Senate Subcommittee on Energy Production and Supply of the Committee on Energy and Natural Resources, October 25, November 15, and November 16, Part I. Washington: U.S. Government Printing Office, 1977.

Hearings on *Leasing and Western Development of Coal* before U.S. Senate Subcommittee on Energy Production and Supply of the Committee on Energy and Natural Resources, February 8 and February 10, 1978, Part II. Washington: U.S. Government Printing Office, 1978.

Leasing Policy Development Office of the Department of Energy. *Federal Coal Leasing and 1985 and 1990 Regional Coal Production Forecasts.* Washington, 1978.

Oversight Hearings on *Federal Coal Leasing* before the U.S. House of Representatives Subcommittee on Mines and Mining of the Committee on Interior and Insular Affairs, March 26, April 1, and April 5, 1976. Washington: U.S. Government Printing Office, 1976.

Sierks, William R., Dean, Norman L., and Thompson, Grant P. *Federal Coal Leasing Policy: A Status Report,* Draft, January 6, 1978. Washington: Environmental Law Institute, 1978.

United States Department of Interior. *Annual Report on Coal for the Fiscal Year 1977.* Washington, 1978.

United States Department of Interior. *Final Environmental Impact Statement: Proposed Federal Coal Leasing Program.* Washington, 1975.

* * * *

Leasing Policy Development Office, Department of Energy, Washington.

Office of Coal Leasing and Planning Coordination, Department of Interior, Washington.

Office of Coal Management, Bureau of Land Management, Department of Interior, Washington.

Western regional offices of the Bureau of Land Management and the Geological Survey.

* * * *

Environmental Policy Institute, Washington.

Natural Resources Defense Council, Washington.

Friends of the Earth, Washington.

Sierra Club, Washington.

STATE LEASING

Council of State Governments. *Coal: State Coal Severance Taxes and Distribution of Revenues.* Lexington, 1976.

North Dakota State Land Department. *Handbook on Western Coal Leasing.* Bismarck, 1976.

INDIAN LEASING

Comptroller General of the United States. *Report to the Senate Committee on Interior and Insular Affairs: Indian Natural Resources, Part II: Coal, Oil, and Gas Better Management Can Improve Development and Increase Indian Income and Employment.* Washington: The General Accounting Office, March, 1976.

Ruffing, Lorraine Turner. *Navajo Mineral Development.* (Unpublished manuscript), 1977.

* * * *

American Indian Policy Review Commission, Washington.

Bureau of Indian Affairs, Department of Interior, Washington.

Crow Coal Authority, Crow Agency, Montana.

Crow Coal Council, Crow Agency, Montana.

Crow Tribal Council, Crow Agency, Montana.

Hopi Tribal Council, Hopi Reservation, Arizona.

Minerals Department of the Navajo Tribe, Window Rock, Arizona.

Navajo Tax Commission, Window Rock, Arizona.

Navajo Tribal Council, Window Rock, Arizona.

Northern Cheyenne Tribal Council, Lame Deer, Montana.

Uintah/Ouray Tribal Council, Uintah and Ouray Reservation, Utah

*   *   *   *

Council of Energy Resource Tribes, Washington.

DNA—Navajo Legal Aid, Window Rock, Arizona.

National Indian Youth Council, Albuquerque, New Mexico.

COLORADO

Colorado Department of Natural Resources. *Summary of Transactions of the State Board of Land Commissioners of Colorado for the Period July 1, 1976 through June 30, 1977.* Denver, 1977.

Colorado Division of Mines. *A Summary of Mineral Industry Activities in Colorado, 1976; Part I: Coal.* Denver, 1976.

*   *   *   *

Bureau of Land Management, U.S. Department of Interior, Denver.

Colorado Board of Land Commissioners, Denver.

Colorado Department of Natural Resources, Denver.

Colorado Department of Revenue, Denver.

Colorado Division of Mines, Denver.

Geological Survey, U.S. Department of Interior, Denver.

*   *   *   *

Citizen's Coal Policy Institute, Denver, Colorado.

Colorado Open Space Council, Inc., Denver, Colorado.

Environmental Policy Institute, Denver, Colorado.

MONTANA

Montana Department of State Lands. *Statistical Report for the Period July 1, 1974 through June 30, 1976.* Helena, 1977.

Montana State Department of Revenue. *Montana Tax Digest.* Helena, 1977.

Montana State Department of Revenue. *Report of the Montana State Department of Revenue to the Legislature for the Period July 1, 1974 through June 30, 1976.* Helena, 1977.

*   *   *   *

Bureau of Indian Affairs, U.S. Department of Interior, Billings.

Bureau of Land Management, U.S. Department of Interior, Billings.

Geological Survey, U.S. Department of Interior, Billings.

Montana Department of Natural Resources and Conservation, Helena.

Montana Department of Revenue, Helena.

Montana Department of State Lands, Helena.

Montana Environmental Quality Control, Helena.

*　　*　　*　　*

Environmental Information Center, Helena.

Friends of the Earth, Billings.

Northern Plains Resource Council, Billings.

Northern Rockies Action Group, Inc., Helena.

NEW MEXICO

Liebendorfer, John. *Public Land Private Profit: A Study of Coal Leasing on State Land in New Mexico.* Albuquerque: Southwest Research and Information Center, 1977.

Governor's Energy Impact Task Force. *Managing the Boom: Socioeconomic Impacts of Energy Development in Northwestern New Mexico,* Final Report, vol. 1. Santa Fe, 1977.

New Mexico Energy Institute. *Coal Industry in New Mexico.* Report Number 76 - 100A. Albuquerque: University of New Mexico, 1976.

New Mexico Legislative Energy Committee. *Legislative Energy Report 2: Taxes and Revenues,* vol. 1. Santa Fe, 1976.

New Mexico State Investment Council. *Nineteenth Annual Report on the State Permanent Funds for the Period July 1, 1976 through June 30, 1977.* Santa Fe, 1977.

New Mexico State Land Office. *Annual Report of the Commissioner of Public Lands for the Period July 1, 1975 through June 30, 1976.* Santa Fe, 1977.

*　　*　　*　　*

Bureau of Land Management, U.S. Department of Interior, Santa Fe.

Geological Survey, U.S. Department of Interior, Carlsbad.

Bureau of Indian Affairs, U.S. Department of Interior, Albuquerque.

New Mexico Bureau of Mines and Mineral Resources, Socorro.

New Mexico Energy Resources Board, Santa Fe.

New Mexico State Investment Council, Santa Fe.

New Mexico State Land Office, Santa Fe.

*　　*　　*　　*

New Mexico Citizens for Clean Air and Water, Los Alamos.

Southwest Research and Information Center, Albuquerque.

NORTH DAKOTA

North Dakota Tax Commission. *Biennial Report for the Period July 1, 1975 through June 30, 1977.* Bismarck, 1977.

Voelkner, Stanley W., Taylor, Fred R., and Ostenson, Thomas K. *The Taxation and Revenue System of State and Local Governments in North Dakota.* Agricultural Economics Report Number 117. Fargo: North Dakota State University, 1976.

*   *   *   *

North Dakota Board of University and School Lands, Bismarck.

North Dakota Coal Development Impact Office, Bismarck.

North Dakota Regional Environmental Assessment Program, Bismarck.

North Dakota State Tax Department, Bismarck.

UTAH

Utah Division of Oil, Gas, and Mining of the Department of Natural Resources. *The Oil and Gas Conservation Act, The Mined Land Reclamation Act, and General Rules and Regulations and Rules of Practice and Procedure.* Salt Lake City, 1977.

Utah Division of State Lands of the Department of Natural Resources. *Fiscal Report for the Period July 1, 1976 through June 30, 1977.* Salt Lake City, 1977.

Utah Geological and Mineral Survey. *Utah Mineral Industry Operator Directory, 1977,* bulletin III. Salt Lake City, 1977.

*   *   *   *

Bureau of Land Management, U.S. Department of Interior, Salt Lake City.

Geological Survey, U.S. Department of Interior, Salt Lake City.

Utah Division of State Lands of the Department of Natural Resources, Salt Lake City.

Utah Geological and Mineral Survey, Salt Lake City.

Utah Tax Commission, Salt Lake City.

*   *   *   *

Environmental Defense Fund, Ogden.